96

WHAT THE QUEEN MEANT TO US

96

WHAT THE QUEEN MEANT TO US

A collection of moments, memories and stories.

ANDREW LANNERD

BROWN DOG BOOKS

First published 2025

Published under licence by Brown Dog Books and
The Self-Publishing Partnership Ltd, 10b Greenway Farm, Bath Rd,
Wick, nr. Bath BS30 5RL, UK

www.selfpublishingpartnership.co.uk

ISBN printed book: 978-1-83952-939-9
ISBN e-book: 978-1-83952-940-5

Cover design by Kevin Rylands
Internal design by Mac Style

Printed and bound in the UK

This book is printed on FSC® certified paper

MIX
Paper | Supporting responsible forestry
FSC® C013604

A tribute to the life, reign and legacy of
Her Majesty Queen Elizabeth II (1926–2022)

Her Majesty Queen Elizabeth II
Edinburgh, Scotland
30 June 2019

Image courtesy of Ken Stone

Contents

Introduction

280,000 people were determined to queue no matter the conditions. 280,000 people were there to witness and take part in a historical moment. 280,000 people felt they had to be there.

These numbers comprise the people who queued for as many as 24 hours to file past the coffin of Queen Elizabeth II, the longest reigning monarch in British history. At its peak the queue stretched for 10 miles across London.

Most of those slowly edging towards Westminster had never known another monarch, and in a way their own lives were intertwined with the milestones of her reign. Finally reaching the top of the ancient staircase in Westminster Hall, many became so struck with the moment that they failed to notice the discreet television cameras there to capture the scene.

As they descended the stone staircase, with their long wait finally over, they stood alongside the coffin. Some bowed their heads, some curtsied and some cried as they came face to face with a significant moment in British history.

Within Westminster Hall many pivotal and historical events in the story of England, Britain and the United Kingdom had taken place – the trial of King Charles I in 1649, the splendour of Coronation banquets and countless other events that have shaped the story of the British Isles.

Under the hall's medieval hammerbeam roof, was the coffin of a monarch, Lying in State. The jewels in the Imperial State Crown, the Sovereign's Orb and Sceptre with Cross sparkled alongside the flickering candles at each corner of the catafalque.

Queen Elizabeth II had inspired, touched and intrigued countless millions around the world, and the faces of the people filing past carried with them the weight of their respect and emotion.

They represented every race, age, culture and creed.

Their diversity was clear, yet they were not there because they were supposed to be there, they were there because The Queen meant something personal to them.

In addition to the many thousands that formed 'the Queue' there are millions of others with a story to tell about how this monarch had touched their own lives.

As I watched the Lying in State, I was deeply struck by the faces of the diverse individuals filing past – filled with emotion, gratitude and respect.

I felt compelled to collect some of their stories.

96 individual moments, memories and stories about what The Queen meant to us.

Andrew Lannerd
London, Spring 2025

8–19 September 2022: A Historical Survey

At 3.10 p.m. on Thursday 8 September 2022, Queen Elizabeth II died at Balmoral Castle 96 years, 141 days, 12 hours and roughly 30 minutes after she was born, and 70 years, 214 days and some 10 hours after she acceded to the throne. Having released a statement earlier that day expressing concern over Her Majesty's health, the Royal Family made the public announcement at 6.30 p.m. confirming that the longest reign of a British monarch – and the second longest of any monarch of a sovereign state in history – had come to an end. Union flags at Buckingham Palace, Hillsborough Castle and 10 Downing Street were lowered to half mast, as were the Royal Banner of Scotland at the Palace of Holyroodhouse and the Welsh flag at Cardiff Castle. At Balmoral, the Royal Standard of the United Kingdom remained raised in acknowledgement of the new King's presence. Britain had now entered a period of national mourning.

As typically happens at times of high royal significance, crowds quickly gathered around Buckingham Palace and other royal residences. Estlyn Davies, Deputy Church

Warden at St John the Baptist in Windsor, noted that as she was closing up on Thursday night roughly 600 people had gathered at the church door, and 400 lit candles in The Queen's memory.* Books of Condolence were opened at cathedrals and in thousands of public buildings across the country, and a dedicated page for tributes was set up on the Royal Family's official website. Operation London Bridge, the protocol to be followed in the event of The Queen's death, and Operation Unicorn, that which would be followed should she die in Scotland, had now been activated, setting in motion an intricate day-by-day plan for the 10-day mourning period. First, The Queen's body would be transported from Balmoral to Holyroodhouse, and the day after to St Giles' Cathedral, Edinburgh, where she would Lie at Rest for 24 hours. From Edinburgh, The Queen would be flown to London where she would rest at Buckingham Palace for the night before being processed to Westminster Hall for a Lying in State of 4 days before the State Funeral at Westminster Abbey. After the funeral, The Queen would travel to Windsor where committal and interment services would be held at St George's Chapel.

The following day, between the hours of 12.00 and 1.00 p.m., the tolling of muffled church bells rang out across the United Kingdom. At 1 o'clock, 71 horses from the King's Troop, Royal Horse Artillery advanced from Wellington Barracks to Hyde Park, 36 of whom were drawing 6 First World War-era QF 13-pounder field guns. The guns were positioned 10 metres apart and directed over Buckingham Palace as the order was given to fire a 96-round gun salute, each round representing a year of The Queen's life. Simultaneous salutes were fired from the Tower of London, as well as authorised stations and British Army and Royal Navy bases across the four nations.

In spite of government guidance stating there was no obligation to do so, many events due to take place over the forthcoming days were immediately postponed as a mark of respect. Audience members at the

* Writing for Premier Christianity.

Mercury Prize ceremony on Thursday evening had already taken their seats when The Queen's death was announced and proceedings were duly called off. The BBC Proms concert on Friday, and the renowned Last Night of the Proms on Saturday, were likewise cancelled. The first Hackney Carnival since the Covid-19 pandemic, scheduled for Sunday, was put off for another year. The National Television Awards, set for 15 September, were delayed until October, and plans for the BBC Radio 2 Live festival, to be held in Leeds on 17 and 18 September, were abandoned altogether. Strikes arranged during the period of national mourning were also suspended by the Rail, Maritime and Transport union. In sport, the Football Associations of England, Scotland, Wales and Northern Ireland decided to suspend all matches over the weekend, from grassroots to top divisions. The British Boxing Board of Control postponed all British boxing over the weekend including Saturday's main event, the women's middleweight world title fight between Savannah Marshall and Claressa Shields at the O2 Arena. Numerous cricket, cycling, Rugby Union and Rugby League, golf, athletics and horseracing events were also cancelled. Departments stores Selfridges and Liberty also closed their doors in the days following The Queen's death, while Fortnum & Mason blacked out its windows, stopped its famous façade clock outside the Piccadilly storefront, and issued all customer-facing teams with black armbands.

News of The Queen's death quickly occupied a solitary position at the centre of the world stage. Google Search data from Thursday shows that 'queen' was the globe's most searched item, and the words 'Queen Elizabeth' in 5 different languages occupied the top 5 places on the trending chart. The Royal Family's announcement on social media platforms had the world's highest number of engagements of any post on 8 September across Instagram, Facebook and Twitter, and all of the world's top 10 most viewed English-language Wikipedia articles were related either to Queen Elizabeth or the Royal Family. In Britain, the BBC had been covering Queen Elizabeth's condition constantly ever since the Palace released its initial statement regarding her health.

Special reports were also running on ITV, Channel 4, Channel 5 and Sky News channels. It has been estimated by British ratings aggregators that on television alone, approximately half of the UK (33,000,000 people) tuned in to the coverage between midday on Thursday and 2.00 a.m. the following morning.

Predictably, interest in travelling to the United Kingdom in order to witness this historical occasion rose. Demand for flights from the United States to London reportedly went up by 49% immediately after the news broke. European countries such as France, Italy, Albania and the Czech Republic were some of the other countries with sizeable spikes in flight searches to London, with Albania as the leader with a 730% increase. National Rail and TfL issued warnings that travel into and throughout London would be extremely congested over the next week as mourners gathered. Many travel agencies declared London to be pretty well full over the period, particularly on the days leading up to the State Funeral, with need for rooms the highest it had been since the London Olympics in 2012. As a result, police, foreign delegations and media had to block book sections of hotels in key areas, meaning the rate at which they reached capacity was even higher.

Hoteliers found it difficult to satisfy the volume of requests. Travelodge, which has 78 hotels across the capital, pronounced all its accommodation in London and Windsor to be completely sold out. Many of the hotels in central London and Windsor closed off their online booking systems, preferring only to deal with enquiries ad hoc, and setting rates at their discretion. Some hotels raised prices by more than 300%. Luxury hotel ME London on the Strand, for instance, was charging a total of £5,028 for two nights – Sunday 18 and Monday 19 September – in a 'Mode' room for two people. The same room booked for a week later – Sunday 25 and Monday 26 September – would cost £1,148. The cheapest room at Park Plaza County Hall London – one of the closest hotels to Westminster Abbey – on Sunday night would cost £1,299, compared with £269 seven days later. It should be noted that of the many thousands travelling to London over the mourning period, more than 60% were from foreign countries, meaning considerable sums were being outlaid to get there.

For most people, though, the opportunity to pay tribute to the only British monarch in the vast majority's living memory was difficult to pass up. Indeed, bearers of flowers had begun to arrive at royal residences on Thursday evening and continued in a steady stream until well after the State Funeral. Dedicated sites for bouquets were established at centres around Britain, with local councils putting out notices with guidance on where flowers should be laid in smaller community areas. Buckingham Palace introduced a system whereby flowers left at the palace gates would be removed after 12 hours to a floral display in Green Park. Cellophane wrapping was prohibited. Between 7 and 14 days after The Queen's funeral, all bouquets would be taken to Hyde Park nursery for processing. Remaining packaging, cards and labels would be recycled and flowers would be sent to Kensington Gardens to be made into compost for use on Royal Parks. These measures were partly taken to avoid a repeat of what happened after the death of Diana, Princess of Wales in 1997, when roughly 60,000,000 flowers were left to rot in their packaging outside Kensington Palace. Early estimates were suggesting those numbers would even be exceeded this time around.

The British Florist Association warned vendors that the market would be significantly inflated over the coming week. Many high street florists sold out completely within a matter of days. Marks & Spencer, Sainsbury's and Tesco stores close to tribute sites and train stations all reported empty stocks of flowers at different points. The florist Darling Buds of Windsor sold more than 70,000 flowers from their shop in the Berkshire town, with sunflowers – a symbol of loyalty and eternal warmth – being among the most popular. Many varieties were made unavailable by wholesale distributors because of diminishing supplies. Flower producers in Turkey were already working hard to fulfil orders in time for the State Funeral. Turkish Cargo declared more than 500,000 flower stems, weighing about 13 metric tonnes, were delivered to the UK for the funeral alone. One wholesale florist in the Netherlands reported that their revenue reached in excess of €140,000 in that week, compared to a normal week of €10–15,000.

Flowers were not the only thing that well-wishers were leaving. Harrods confirmed that its 42cm soft toy of Paddington Bear – with whom The Queen had appeared in a popular sketch produced for her Platinum Jubilee concert – was out of stock. By 12 September, the Royal Parks had issued a statement discouraging mourners from leaving marmalade sandwiches – Paddington's favourite food – in floral tribute areas. But with sales of marmalade up by more than 40% at certain retailers, it appears as though many were content to honour Her Majesty's memory at home. Other items followed suit. The wine-based liqueur Dubonnet – known to have been Queen Elizabeth's favourite tipple – sold out across Britain, and increased by 465% in sales in Australia. The prices of corgi puppies, the dog breed with which The Queen had come to be so closely associated, climbed to more than £6,000 in some places, more than five times their usual rate, and searches for the dogs on pet rehoming websites increased tenfold. Google searches for 'Royal memorabilia' also doubled over the weekend. The Royal Mint website crashed within hours of The Queen's death after customers rushed to buy commemorative coins. At one point there were more than 6,390 people queuing to make their orders. By 11 o'clock on Friday morning, several Elizabeth II coins had already been listed as unavailable, from a florin from 1967, priced at £8, to a 2005 gold proof crown marking the 200th anniversary of the Battle of Trafalgar, costing £3,575.

All the while, Operation London Bridge and Operation Unicorn were in full swing. On Sunday, focus was again on Balmoral Castle as The Queen left her beloved summer residence for the very last time. Her coffin, made 30 years earlier from British oak and, following a royal custom dating back to Elizabeth I, lined with lead, weighed around a quarter of a tonne. It was draped with the Royal Standard of

Scotland and topped with a wreath of flowers from the Castle gardens. With thousands lining the route, the royal cortège made the 175-mile journey through Aberdeenshire, Aberdeen, Angus, Dundee, Perth and Kinross, and Fife, before arriving in Edinburgh at Holyroodhouse in the afternoon. Elizabeth was the only British monarch ever to die in Scotland and when her body was transported through the streets of Edinburgh, it was the first time that a royal cortège had passed through the Scottish city since the burial of James V in 1543. In a military procession that included King Charles, Princess Anne and Sir Timothy Laurence, Prince Andrew, Prince Edward, the Bearer Party from the Royal Regiment of Scotland and the Royal Company of Archers, The Queen was carried the 1,200 yards up the Royal Mile to St Giles' Cathedral where she would Lie at Rest for 24 hours with the Crown of Scotland placed on her coffin. Guns were fired every minute from Edinburgh Castle as the journey was made.

On Sunday evening, queues of people waiting to file past the coffin had begun forming in the Meadows to the south of the cathedral, with many people preparing to wait through the night. That night, a thanksgiving service in celebration of The Queen's life was held in the cathedral. It was attended by the royal party, British Prime Minister Liz Truss and the First Minister of Scotland Nicola Sturgeon, as well as representatives from The Queen's Scottish charities and organisations. Next morning, the back of the queue had extended to more than a mile from St Giles'. The first person to have joined in line, Sharon O'Neill, had travelled from Airdrie. Speaking afterwards, she said, 'It was stunning, it really was, but also very sad, and a very poignant moment.' The last wristbands were handed out at 1.00 p.m., and the cathedral doors were closed at 3.00 p.m. Roughly 33,000 people went through, with the longest wait time said to have been 8 hours and 20 minutes from the time of doors opening. The last two people allowed to join the queue were Charlie, 68, and his mother Janet, 89. The mother and son had made the journey from Cathcart, Glasgow. Charlie told reporters he was 'very honoured' to be in

Edinburgh for this 'very sombre occasion', adding that it was important he could take his mother, whom he was pushing in a wheelchair, to be there. Janet got married in 1952 and bought her first television set the following year especially to watch The Queen's Coronation.

After leaving St Giles' on Tuesday morning, The Queen's coffin was taken by hearse to Edinburgh Airport, attended by Princess Anne and Sir Timothy Laurence, whence she would fly to RAF Northolt. The Royal Air Force Bearer Party conveyed the coffin on to the C-17 Globemaster and a Guard of Honour was formed by the Royal Regiment of Scotland. 6,000,000 people tried to track the flight's progress on Flightradar24 within a minute of it being uploaded, causing the website to crash. A further 396,000 watched on a YouTube live stream. The journey lasted 1 hour and 12 minutes, during which time the Scottish Royal Standard draping the coffin was replaced by the Royal Standard of the United Kingdom. From Northolt, a cortège travelled to Buckingham Palace for the night before making its way to Westminster Hall for the four days' Lying in State before the funeral. Once inside the Hall, the Imperial State Crown, an ornament adorned with some 2,868 diamonds, 17 sapphires, 11 emeralds, 269 pearls and 4 rubies, was placed on top of the coffin.

Preparations for the queue of mourners at Westminster had long been underway. A security checkpoint was installed before the entrance, and airport-style barriers were set up in Victoria Tower Gardens. The line would cross Lambeth Bridge and stretch eastwards along the southern embankment of the Thames. 500 portable toilets, 8 first aid stations, and manifold water stations were posted along the route. Additional trains and buses were timetabled to ensure visitors were able to return home, and King's Cross and Victoria Stations were kept open 24 hours a day to offer shelter to those waiting for transport.

Doors to Westminster Hall would open at 5 p.m. on Wednesday, but the numbers began to gather well in advance. The first person to join the queue was 56-year-old Vanessa Nathakumaran from Harrow, who arrived at Lambeth Bridge at 12 p.m. on Monday – a full 2 and a half days before the queue would open – with enough provisions to spend the duration there. Speaking to the press, Nathakumaran believed it was her 'duty' to be first in line to file past The Queen, and the least she could do in return for the late monarch's decades of service. By Monday evening many more had joined and the queue progressively lengthened until, when the doors were opened on Wednesday, more than 100,000 were already waiting. The Greater London Authority and the Department of Culture, Media and Sport (DCMS) coordinated operations, with in-person support provided by approximately 800 paid stewards, assisted by 100 civil service volunteers and more than 300 Scouts. Volunteers from non-profit groups such as the British Red Cross, Samaritans and Salvation Army, together with aides from St John Ambulance and First Aid Nursing Yeomanry were also in attendance. Mourners were issued with coloured wristbands upon entering the queue, so that they could keep their place should they need to leave temporarily. The DCMS provided live updates on the status of the queue on Twitter, Instagram and YouTube, with progress being closely followed by the mainstream media.

The longest waiting time for the queue surpassed the 24-hour mark, despite authorities attempting to suspend new entrants once 14-hour waits had been reported. At its maximum length, the queue reached all the way to Southwark Park, where a zigzag formation developed, making its total length in excess of 10 miles. It has been estimated that as many as 280,000 people filed through Westminster Hall, with so much continuous footfall causing the Yorkstone floor inside the Hall to be delaminated. The London Ambulance service stated that its staff and volunteers provided medical aid to an estimated 2,000 people in the queue, 240 of whom were taken to hospital for treatment. The last person to receive a wristband was Chrissy Heerey, a serving member of the RAF at Melton Mowbray, who was joining the queue for a second time. Heerey entered the Hall at 6.30 a.m. on Monday.

Only a few hours later, at 10.44 a.m., the State Gun Carriage of the Royal Navy moved The Queen's coffin from Westminster Hall to Westminster Abbey in preparation for the State Funeral. The carriage was drawn by 98 Royal Navy sailors, known as Naval Ratings, with 40 more marching behind as brakes, keeping up a tradition that began at the funeral of Queen Victoria. King Charles, members of the Royal Family, and members of The Royal Household followed behind the procession. A wreath of flowers cut from the gardens of Buckingham Palace, Highgrove House and Clarence House was placed on the coffin, together with a note from The King which read 'In loving and devoted memory. Charles R.' The carriage arrived at the Abbey at 10.52 a.m. The last time a British monarch's State Funeral was held in the Abbey was King George II's in 1760.

The service would be attended by 2,000 guests, 500 of whom were foreign dignitaries and heads of state, thus making it the largest gathering of world leaders in recorded history. Invitations had been handled by the Foreign Office, alongside communications and security arrangements for individuals, all of which was organised from a headquarters called 'The Hangar', which redeployed 300 staff to oversee the process. Representatives of every country with which Britain maintains diplomatic relations were invited, excluding Russia, Belarus, Afghanistan, Myanmar, Venezuela and Syria. The programming of the funeral allowed those intending to address the UN General Assembly at its general debate the following day enough time to travel to New York City.

London transport authorities forecast 1,000,000 people would travel to the capital on Monday, with 250 extra rail services running to accommodate the increase in traffic. Many were expected to fill the streets around Westminster and Buckingham Palace. With such high-profile guests in the congregation at Westminster Abbey, the level of security required would be unprecedented. Stuart Cundy, Deputy Assistant Commissioner for the Metropolitan Police, said The Queen's funeral would be the Met's largest ever single deployment of police officers in London, bigger than the 2012 Olympics and the Platinum Jubilee weekend. At least 10,000 officers and 1,500 military personnel would be on duty, and 22 miles of barriers would be assembled in central London to help marshal the crowds. A Special Operations Room was launched in Lambeth, where representatives of the Met, GCHQ, MI5, emergency services and local authorities joined forces to orchestrate the intricate security and surveillance operation. Should the SOR become for some reason inoperable, such as in the event of a power outage, a backup centre was ready 10 miles away. Hundreds of police spotters and snipers were deployed on the tops of buildings around central London, and the Met had 3 helicopters monitoring the skies. Every manhole cover and lamp post was checked for explosives.

Before the service began at 11.00 a.m. the Abbey's tenor bell was tolled every minute for 96 minutes in reflection of the years of Queen Elizabeth's life. Flights scheduled to fly over London were diverted to minimise noise and as a mark of respect. The service was conducted according to the 1662 Book of Common Prayer by the Very Reverend Dr David Hoyle, Dean of Westminster. Singing was led by the Choir of Westminster Abbey and the Choir of the Chapel Royal, St James's Palace, under the direction of James O'Donnell. As the coffin entered the Abbey, the choir sang William Croft's funeral sentences. The lessons were read by Baroness Scotland, Secretary General of the Commonwealth, and Liz Truss, and the sermon and commendation were given by the Archbishop of Canterbury, Justin Welby. Prayers were recited by clergy from several Christian denominations. A psalm setting by Judith Weir and an anthem by James MacMillan were both written for the funeral. At the end of the service there was a sounding of the 'Last Post', followed by a 2-minute silence, concluded with the '*Reveille*'. A rendition of 'God Save the King', and the lament for bagpipes 'Sleep, Dearie, Sleep' closed the ceremony. The Queen's coffin was carried out to Bach's *Fantasia in C minor*. A procession of 2,650 military personnel took The Queen's coffin to Wellington Arch, and from there it would be taken by State Hearse to Windsor for the committal service at St George's Chapel.

An estimated 215,000 people had gathered around Buckingham Palace and Westminster for the funeral, with 75,000 in ceremonial viewing areas, 60,000 on South Carriage Drive and 80,000 in Hyde Park. Hundreds of thousands more were in central London and lining the route out of London to Windsor. The day had been declared a bank holiday in the UK, with the London Stock Exchange, major supermarkets and many businesses, workplaces and education centres having agreed to close. Several NHS trusts chose to reschedule or cancel non-urgent appointments. The Queen's funeral and committal services were the first of any British monarch to be broadcast live on national television. 125 cinemas across the country opened their doors to show it. Millions more watched at home. According to the National Grid, a 2 gigawatt power drop – the equivalent of 200,000,000 light bulbs being switched off – was recorded as people across the nation stopped what they were doing to watch the funeral. Energy usage gradually increased by a further 500 megawatts over the course of the ceremony, as people began to resume normal activities such as putting on the kettle. Those levels plummeted again by 500 megawatts during the 2-minute silence. Global ratings of the funeral have been estimated at around 4,000,000,000 viewers, making it the largest event in televised history. The Queen was once again the most trending Google search topic on that day.

When the State Hearse arrived at Windsor, a final procession of the State Gun Carriage went to St George's Chapel. 410 military personnel took part, while 480 lined the streets, 150 stood in a guard of honour and lined the steps, and a further 130 were fulfilling other duties. Citizens wishing to attend the procession in person needed to go through security checks, and 2,000 police officers were monitoring the town. Also in attendance during the procession were 3 of The Queen's remaining pets: her last 2 corgis, Muick and Sandy, and Emma the Fell pony, who had Queen Elizabeth's riding headscarf lain across the saddle. The Sebastopol Bell and the Curfew Tower bell tolled, and the King's Troop, Royal Horse Artillery fired minute guns from the East Lawn of Windsor Castle. The coffin entered St George's Chapel via the West Steps. The committal service, conducted by the Right Reverend David Conner, Dean of Windsor, was at 4.00 p.m. The congregation was largely made up of Royal Household and staff from The Queen's private estates, in addition to Royal Family members and leaders from Commonwealth realms. Towards the end of the service the Imperial State Crown, orb and sceptre were removed from the coffin and positioned on the altar. King Charles placed the Queen's Company Camp Colour of the Grenadier Guards on the coffin, before the Lord Chamberlain broke his wand of office and placed its halves upon the coffin. The Garter Principal King of Arms recited the styles of Elizabeth II and Charles III, between which a lament – 'A Salute to the Royal Fendersmith' – was played by the Sovereign's Piper as The Queen's coffin was lowered into the Royal Vault. The singing of the National Anthem marked the end of the ceremony.

The sense of timelessness and familiarity with which Queen Elizabeth's subjects had come to associate her meant that the significance of her absence was not yet fully formed in British consciousness. During her reign The Queen was served by 15 British prime ministers, gave her assent to more than 4,000 Acts of Parliament, travelled over 1,000,000 miles in official visits, and sent more than 300,000 telegrams or cards to her subjects celebrating a 100th birthday, or 900,000 to those marking a Diamond Wedding Anniversary. She is the single most photographed person in history, and her likeness is the world's most reproduced, with 35 different countries having produced billions of coins and stamps with her effigy. The Machin portrait – one of 6 moulds of Elizabeth's profile created by Stoke-on-Trent artist Arnold Machin, subsequently photographed and turned into the iconic portrait found on the majority of British stamps since 1967 – is the most reproduced image in human history, having been reprinted well over 300,000,000,000 times. With such a longstanding and ubiquitous presence no longer felt, many people were reminded of personal griefs. The British Association for Counselling and Psychotherapy and the Samaritans each published statements warning that times of national mourning can affect people in different ways. Church attendance was also seen to increase as the need for unity – and stability – rose in Britons. The nation was waking up to a new era of the British monarchy, yet its public had no precedent for what exactly that might look like.

Cyril Worthington-Feckenham

96

Facts About the Life, Legacy and Reign of Queen Elizabeth II

1

Princess Elizabeth (later to become Queen Elizabeth II) was born at 17 Bruton Street in Mayfair (London) on 21 April 1926. The royal birth made headline news throughout the world.

2

Her parents were Prince Albert, Duke of York (who later took his father's name when he became King George VI) and his wife, Lady Elizabeth Bowes-Lyon (later Queen Elizabeth, and then Queen Elizabeth the Queen Mother after 6 February 1952).

3

Her full name was Elizabeth Alexandra Mary Windsor.

4

At the time of her birth, Princess Elizabeth was not expected to become Queen, as her uncle was the heir to the throne (later to become King Edward VIII and then Duke of Windsor).

5

She spent the early years of her life in London with her family at 145 Piccadilly, their London residence (later destroyed during The Blitz). She could see Buckingham Palace across Green Park from the nursery windows.

6

Her grandfather – King George V – was monarch from 1910 to 1936. Princess Elizabeth referred to her him as 'Grandpa England'.

7

She was christened on 29 May 1926 in the private chapel at Buckingham Palace with water brought from the River Jordan by the Archbishop of York, Cosmo Gordon Lang. Her godparents were: Lady Elphinstone (her maternal aunt); Prince Arthur, Duke of Connaught (her great-great-uncle); the Earl of Strathmore (her maternal grandfather); and Princess Mary (her paternal aunt).

8

Her nickname amongst the family was Lilibet, which originated during her childhood and continued to be used until her death.

9

Princess Elizabeth's cousin Margaret Rhodes once said that she was 'a jolly little girl, but fundamentally sensible and well behaved'.

10

In 1928 Winston Churchill described Princess Elizabeth in a letter to his wife Clementine, saying: 'She has an air of authority and reflectiveness – astonishing in an infant.' He would later become her first prime minister.

11

She loved horses from a very young age and was given Peggy, a Shetland pony, for her fourth birthday. Horses remained a significant passion throughout her life.

12

In 1930, her younger sister Princess Margaret was born at Glamis Castle in Scotland. Despite very different personalities they would remain close throughout their lives. Princess Margaret died in 2002 at the age of 71.

13

In 1934 at the age of 8, Princess Elizabeth was introduced to her distant cousin and future husband, Prince Philip of Greece and Denmark. Philip was five years older, born on 10 June 1921 at *Mon Repos* on the Greek island of Corfu.

14

Her grandfather King George V died on 20 January 1936 at Sandringham, and his eldest son Edward – Elizabeth's uncle – became King Edward VIII.

15

On 11 December 1936, King Edward VIII abdicated the throne to marry Wallis Simpson, a twice-divorced American. After a constitutional crisis, Edward renounced the throne in favour of his brother, Albert – Elizabeth's father. These events would have a profound effect on Princess Elizabeth's destiny.

16

Albert adopted his father's name to become King George VI, and moved his family into Buckingham Palace – the official London residence of the British sovereign. Princess Elizabeth wasn't pleased about moving into the draughty palace and complained to her governess.

17

On 12 May 1937, King George VI was crowned in Westminster Abbey. Then 11-year-old Princess Elizabeth attended the ceremony and wrote an essay entitled, *To Mummy And Papa, In Memory Of Their Coronation. From Lilibet, By Herself.* 'I thought it all very, very wonderful and I expect the Abbey did too,' she wrote. 'The arches and beams at the top were covered with a sort of haze of wonder as Papa was crowned, at least I thought so.'

18

The Second World War commenced in September 1939 and many children were evacuated from London to safety in the countryside or abroad. The Royal Family decided against sending Princesses Elizabeth and Margaret away. Queen Elizabeth (later the Queen Mother) famously said at the time: 'The children won't go without me. I won't leave The King. And The King will never leave.'

19

Princess Elizabeth was 13 years old when the Second World War broke out and 19 when it ended. She spent most of the war secretly at Windsor Castle with her sister.

20

On 14 October 1940, Princess Elizabeth gave her first broadcast during the *Children's Hour* radio programme at the age of 14. In it, she sent her best wishes to children who had been evacuated from Britain to keep them safe from the conflict. 'And when peace comes, remember it will be for us, the children of today, to make the world of tomorrow a better and happier place.'

21

During The Blitz of 1940–41, Buckingham Palace was hit 9 times by enemy bombing. Throughout the war Princess Elizabeth learned from her parents about the importance of duty and accompanied them on visits to troops, ships and airfields.

22

After her father succeeded to the throne in 1936, Princess Elizabeth began studying constitutional history and law. She also studied French, German and music and was educated at home.

23

Princess Elizabeth joined the Auxiliary Territorial Service (the women's branch of the British Army during the Second World War) in 1945, becoming the first female member of the Royal Family to join the armed services as a full-time, active member.

24

During her ATS training, she was known as Second Subaltern Elizabeth Windsor. Princess Elizabeth trained as a mechanic and truck driver and graduated as a fully qualified driver just before the war ended. At the time, the press dubbed her 'Princess Auto Mechanic'.

25

When the war ended, huge crowds gathered outside Buckingham Palace to celebrate Victory in Europe Day. Princess Elizabeth and Princess Margaret were permitted to go out and join the crowds in secret, blending in with the public. It is said that Elizabeth was worried about being recognised, but later said it was one of the most memorable nights of her life.

26

During a royal tour of South Africa with her parents and sister in 1947, Princess Elizabeth gave a radio broadcast. In it she famously said: 'I declare before you all that my whole life, whether it be long or short, shall be devoted to your service and the service of our great imperial family to which we all belong.'

27

When Princess Elizabeth was 21, her engagement to Lieutenant Philip Mountbatten was officially announced. Their wedding took place on 20 November 1947 in Westminster Abbey. Winston Churchill summed up the occasion as 'a flash of colour on the hard road we travel'.

28

The royal couple received more than 2,500 wedding presents from all over the world and 10,000 telegrams of congratulations. Many of the wedding presents were put on public display in the State Apartments at St James's Palace in London.

29

The couple's first child, Prince Charles Philip Arthur George, was born on 14 November 1948 at Buckingham Palace. He was later to become, Prince of Wales (1958) and King Charles III (2022).

30

Elizabeth and Philip's second child, Princess Anne Elizabeth Alice Louise, was born at Clarence House on 15 August 1950. She would be given the title Princess Royal in June 1987, one of only seven holders of the title in British history.

31

In the early 1950s, The King's health was in decline. On 6 February 1952, King George VI passed away peacefully in his sleep at Sandringham House in Norfolk.

32

Princess Elizabeth heard of her accession to the throne while on a tour in Kenya *en route* to Australia and New Zealand. She immediately returned to London as Queen Elizabeth II.

33

At the age of 25, Queen Elizabeth II ascended the throne. Her predecessor Queen Elizabeth I also ascended the throne at the age of 25 in 1558.

34

The Queen's first prime minister was Winston Churchill who famously said at her accession: 'Famous have been the reigns of our queens. Some of the greatest periods in our history have unfolded under their sceptres.'

35

On 2 June 1953, Queen Elizabeth II became the 39th sovereign to be crowned at Westminster Abbey.

36

The Queen's Coronation service lasted almost three hours and 8,251 guests attended in Westminster Abbey.

37

The 1953 Coronation made television history, as it was the first ever to be broadcast. 27 million people in the United Kingdom alone tuned in to watch this momentous occasion.

38

The Queen's Coronation dress was designed by Norman Hartnell. It was made of white satin and embroidered with the national symbols of the United Kingdom and the Commonwealth in gold and silver thread.

39

There were more than 2,000 journalists and 500 photographers from 92 nations along the Coronation processional route in 1953.

40

Just under 30,000 people took part in The Queen's Coronation procession on 2 June 1953.

41

The Queen's reign commenced with her longest ever Commonwealth tour between November 1953 to May 1954. The tour encompassed the West Indies, Australasia, Asia and Africa, and covered 44,000 miles.

42

Much of the 1953/1954 royal tour was undertaken on the new Royal Yacht *Britannia*. The Royal Yacht would go on to serve The Queen and her family for a further 44 years, travelling more than one million nautical miles.

43

The Queen and Duke of Edinburgh celebrated the birth of their third child, Prince Andrew Albert Christian Edward, on 19 February 1960. He was born at Buckingham Palace, and was the first baby born to a reigning British Sovereign since the birth of Princess Beatrice to Queen Victoria in 1857.

44

Their fourth and final child, Prince Edward Antony Richard Louis, was born on 10 March 1964. Edward would later become Duke of Edinburgh after the death of his parents. In 1964, four royal births took place within the space of a few months. Princess Alexandra gave birth to James Ogilvy on 29 February, followed by the birth of Prince Edward to Queen Elizabeth II on 10 March, the Duchess of Kent had Lady Helen Windsor on 28 April and then Lady Sarah Armstrong-Jones was born to Princess Margaret on 1 May.

45

The Queen held many different titles, including Head of State, Head of the Nation, Head of the Commonwealth, Head of the Armed Forces, Defender of the Faith and Supreme Governor of the Church of England – and that's to name just a few!

46

As Head of State, The Queen was politically neutral. She served as queen of all of her people. Her diplomatic skills in this area were celebrated on many occasions by her British prime ministers in both political parties and by prime ministers in the Commonwealth Realms.

47

Throughout her long reign, The Queen carried out royal engagements all over the world. In fact, she travelled more than any other monarch, visiting 116 countries.

48

The Queen is credited with inventing the royal 'walkabout' in the 1970s during a tour of New Zealand, where she met local people. The 'walkabout' has now become a very popular royal tradition.

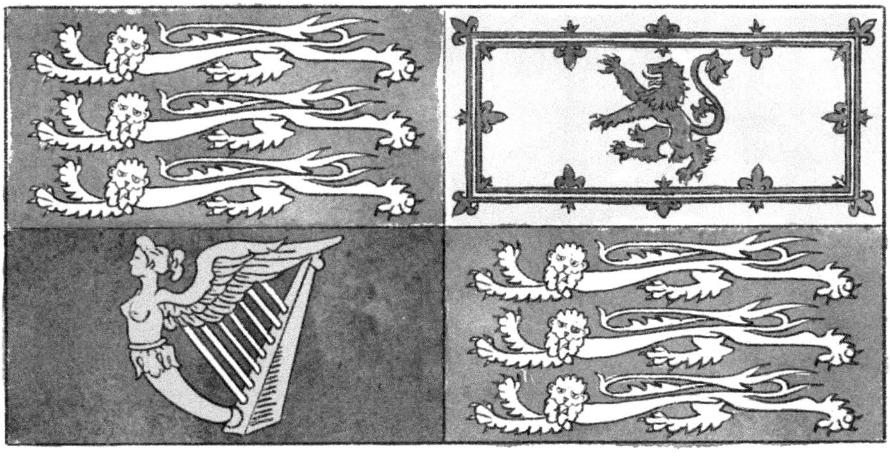

49

While carrying out her duties, The Queen was often given presents from foreign heads of state or well-wishers. These have included horses, but some gifts are a little more unusual. In 1972 when in the Seychelles, she was given two tortoises, while in the same year the President of Cameroon gave her a seven-year-old bull elephant called Jumbo. Many of these presents ended up in the London Zoo.

50

The Queen opened the British Parliament in person on 67 occasions during her reign. There were three occasions when she was unable to do this – 1959 and 1963 when she was expecting Prince Andrew and Prince Edward, and in 2022 during her Platinum Jubilee year when Parliament was opened by her son, The Prince of Wales. The Queen also opened many Commonwealth Parliaments on royal tours over the years.

51

The Queen had links with or patronages of 600 charities and organisations.

52

The Queen visited Canada more than any other country during her remarkable 70-year reign. Her Majesty made a total of 27 visits to the North American nation – a valued member of the Commonwealth.

53

Every year at 3:00 p.m. on Christmas Day, The Queen addressed the nation and the Commonwealth in her annual Christmas speech. The first of these to be broadcast on television was in 1957.

54

The Christmas speech started as a live broadcast. But since 1960, it was recorded by her in advance.

55

In 2012, her annual Christmas speech was broadcast in 3D.

56

Queen Elizabeth II was one of the most famous people ever to have lived. In 2017, Royal Mail estimated her portrait had been reprinted more than 220 billion times around the world in a variety of colours and denominations. In addition to stamps, the image of The Queen was used on billions of coins and banknotes. No human image in the history of the world has been reproduced as much as the image of Queen Elizabeth II.

57

During her record-breaking reign, 15 British prime ministers held office: Winston Churchill, Anthony Eden, Harold Macmillan, Alec Douglas-Home, Harold Wilson, Edward Heath, James Callaghan, Margaret Thatcher, John Major, Tony Blair, Gordon Brown, David Cameron, Theresa May, Boris Johnson, and finally Liz Truss, who was appointed as prime minister just two days before The Queen's death.

58

There were 14 US presidents during The Queen's reign. She met each one, except President Lyndon Johnson.

59

In an average year, The Queen received around 60,000 pieces of printed correspondence. This would rise during anniversaries and Jubilees – for example, the total rose to over 130,000 letters and messages in The Queen's Diamond Jubilee year in 2012.

60

Throughout her reign, The Queen would spend most of the working week at Buckingham Palace. This is typically where she would hold her weekly audiences with the prime minister.

61

Despite being the most famous face the world has ever known, The Queen never granted an interview, and her personal viewpoints where rarely known outside royal circles.

62

Most private weekends were spent at Windsor Castle in Berkshire. When she was in residence at Windsor, she would regularly attend Sunday morning services at the Royal Chapel of All Saints in Windsor Great Park.

63

The Queen typically spent August and September at Balmoral Castle in Aberdeenshire, and the Christmas and New Year period at Sandringham House in Norfolk.

64

In Scotland, The Queen's official residence was the Palace of Holyroodhouse in Edinburgh. Throughout her reign The Queen would spend a week there in late June or early July during a tradition called 'Holyrood Week'.

65

In 2007, The Queen and the Duke of Edinburgh became the first royal couple to celebrate a diamond wedding anniversary (60 years of marriage).

66

The Queen once said about The Duke of Edinburgh: 'He has quite simply, been my strength and stay all these years.' They would celebrate their final and 73rd wedding anniversary on 20 November 2020.

67

Queen Elizabeth II was 5ft 4ins tall. Her predecessor Queen Victoria was 4ft 11ins tall.

68

The Queen was the first British monarch to address the US Congress, which she did during her State Visit to the United States in 1991.

69

The Queen was a passionate lover of corgis. She owned at least 30 during her reign. The corgi became an integral part of her iconography as Queen.

70

The first corgi she owned was named Susan and was a present for her 18th birthday in 1944. Susan was a constant companion and even accompanied Elizabeth and Philip on their honeymoon.

71

By hereditary right, The Queen owned a third of the mute swans on certain parts of the River Thames near Windsor. The annual census of the swan population on this stretch of the River Thames was taken on behalf of The Queen during the ceremony known as 'Swan Upping'. On 20 July 2009 as 'Seigneur of the Swans' Queen Elizabeth II attended the Swan Upping ceremony, the first time a reigning monarch had done so.

72

One of her favourite flowers was the lily of the valley, which was featured in her Coronation and wedding bouquets.

73

The Queen was fond of Scottish country dancing. Every year at Balmoral, she hosted the Ghillies Ball for her estate workers, neighbours and guests. Her Majesty regularly took part in the dancing.

74

In 2014, The Queen sent her first tweet at the Science Museum in London. It said: 'It is a pleasure to open the Information Age exhibition today at the @ScienceMuseum and I hope people will enjoy visiting. Elizabeth R.'

75

During her reign The Queen was the only person in the United Kingdom who did not require a driving licence or a passport. Both driving licences and passports were issued in her name, so she didn't need either document.

76

The Queen was fluent in French, beginning lessons from a young age.

77

The Queen sat for over 200 official portraits during her reign.

78

The Queen had four children, eight grandchildren, and at the time of her death, 12 great-grandchildren.

79

The Queen was the Head of the Commonwealth, a friendly association of 56 countries around the world. Keeping the Commonwealth growing, vital and relevant was one of her great achievements.

80

The Queen made over 150 Commonwealth visits during her reign as Head of the Commonwealth.

81

Her Majesty carried out more than 21,000 engagements over the course of her reign.

82

The Queen gave Royal Assent to approximately 4,000 Acts of Parliament.

83

The Queen hosted 113 incoming State Visits to the United Kingdom. The first incoming State Visit in 1954 was by King Gustaf VI Adolf and Queen Louise of Sweden (who was an aunt of Prince Philip). The last State Visit was in June 2019 when President Donald Trump and First Lady Melania Trump of the United States visited Buckingham Palace.

84

The Queen christened the Victoria Line in 1969, making her the first reigning monarch to travel on the London Underground. She opened the new Elizabeth Line on 17 May 2022 during one of her final public engagements.

85

The Queen was a committed Christian and often spoke of her faith. She once said: 'For me, the life of Jesus Christ, the Prince of Peace, whose birth we celebrate today, is an inspiration and an anchor in my life.'

86

More than 180 garden parties were held at Buckingham Palace, and over 1.5 million people attended a garden party during The Queen's reign.

87

The Queen made a State Visit to Ireland in 2011, the first time a reigning British monarch had visited the country since its separation from the United Kingdom. The visit was a turning point in improving relations between Ireland and the United Kingdom.

88

The Queen conducted over 650 investitures throughout her reign to award citizens with honours for public service.

89

The United Kingdom celebrated a remarkable four jubilees during The Queen's reign – Silver in 1977, Golden in 2002, Diamond in 2012 and Platinum in 2022.

90

The Queen was the custodian of The Royal Collection, held in trust for her successors and for the nation. The Royal Collection is estimated to have over one million individual works of art, including over 8,000 paintings, 160,000 watercolours, prints and drawings, 200,000 photographs, 300,000 items of decorative art and 200,000 books and manuscripts.

91

Queen Elizabeth II reigned longer than any other monarch in British history: 70 years and 214 days.

92

The Queen was the oldest British monarch. The previous holder of this record was Queen Victoria (Elizabeth's great-great grandmother), who died at the age of 81.

93

Not only that, but The Queen was the longest reigning female monarch in world history.

94

At the time of her death, nine out of 10 people across the world had never known the United Kingdom to have a monarch other than Queen Elizabeth II. It was estimated that a third of the UK's population said they either met the Queen or observed her in person, according to YouGov.

95

She was Queen Regnant of 32 sovereign states over the course of her lifetime and remained the monarch of 15 realms at the time of her death.

96

On the day of The Queen's funeral, flights over central London were diverted to ensure that the skies were silent as a mark of respect.

96

Moments, Memories and Stories

COLLECTIONS
ER
POST OFFICE

1

Emma Sommerville
Cheshire, England

'I was actually sitting in the house with the kids when it came on the news and was surprised how upset we all became. We may never have met The Queen but she was Great Britain to us. She was like an unofficial grandmother to everyone, and she will be sorely missed.'

2

Graham McAulay-Sillars
Dundee, Scotland

'I wasn't in the queue to pay my respects to The Queen, but my husband Duncan, our friend Iain and I along with our collective of four corgis waited on the side of the Kingsway (a bypass/motorway in Dundee) for hours to see The Queen's hearse drive past. There were literally hundreds of others doing the same thing. It was such a humbling experience, and I know this will sound maybe a little odd but it's a memory I'll cherish for the rest of my life. I never met The Queen when she was alive and I found it bittersweet that the only time I came close was when she was in her coffin being transported to Edinburgh. The drive-by lasted all of ten seconds but it was truly humbling to know that Her Majesty had passed away in a place she loved, and that place just happened to be the country where I was born.

A place that I love dearly too.

3

Hannah Catherall
Manchester, England

'I've always loved the Royal Family and we grew up watching all of the various celebrations on television, so the day we lost our Queen it felt like part of a family tradition had also died with her.'

4

Sarah Evans
Shropshire, England

'I was actually away celebrating my wedding anniversary. The hotel bar became very subdued when the news was on. The passing of Queen Elizabeth II became expected as her ill health was reported, however no less sad. Her reign and what she achieved was extraordinary and will always be remembered.'

5

Adam Kerr
Worcester, England

'I run a toy shop and we had plans in place for when the day would sadly come. I had posters for the windows and a book of condolence ready. The strange thing was that very morning I got them out to check that there wasn't any damage to them.

I was working away when we received an emergency email to say what had happened. At that point we were told to close our doors straight away and to go home. I live alone and going from such a fun environment where we make children happy with the items we sell, to going home watching the news over and over again was unsettling. I didn't think I would feel that way. We were told that the store would not open the next day but I had to go in to put the posters of The Queen in the window. I've never been a person who followed or took an interest in The Queen but I felt so sad. Putting these posters up in the window by myself with everyone outside watching was so upsetting.'

6

Eloise
Nottingham, England

'I remember early 2022 when my then 3-year-old was at nursery, learning about the Platinum Jubilee, and how he became very interested in The Queen and Royal Family. I learned of The Queen's passing from watching the news, and it struck my child with a lot of sadness.

My family have some fond memories of celebrations and meeting The Queen. My husband first met her during a visit to his school, which had been opened by George VI. He also met her during cadet ceremonies. My in-laws, who live and work near to Sandringham, have a lot of pictures and fond memories from attending the annual garden parties there.'

7

Kate Barton
Harrogate, England

'During the celebrations of The Late Queen's Platinum Jubilee, pupils at Western Primary School, Harrogate, drew pictures and wrote letters to The Queen. These communications, which congratulated her on the longevity and success of her reign as well as telling her a little about themselves and asking questions, were sent to Buckingham Palace. In a twist of fate, a reply to these letters was received from Balmoral Castle

– composed by The Queen's Lady-in-Waiting on Her Majesty's behalf – on 8 September 2022, the day of her death. How poignant that when the children came together for an assembly to acknowledge The Queen's passing the following day, the head teacher was able to use the letter and the beautiful photos that we received to remind the children that, while it was OK to be sad, her life was one to celebrate.'

8

Charlotte
Berkshire, England

'I walked into my mum's house and she had the news on. No announcement had been made at that point but it really wasn't looking good. Then I remember noticing that the newsreader had changed into a black suit . . . that's when we knew. It was confirmed a few minutes later that our Queen had died. My mum was HEARTBROKEN. We decided to make our way to nearby Windsor Castle. Just to be there, lay some flowers and show our respect. The town was HEAVING with people wanting to pay their respects – I've never been there and it be so quiet and sombre. The atmosphere was like nothing I'd felt before; I'll never forget it. Everyone was just so sad.

The Queen will always be a beacon of light and kindness for me. She was pure class and she will forever be my favourite royal.'

9

Nina Smith
Norfolk, England

'I have always loved our Royal Family, followed their lives and life events and celebrations.

After the sad death of Prince Philip and the heartbreaking sight of Her Majesty alone at his funeral, I felt very protective of her.

I was working that day and had put a new display in my shop window. It was filled with fabrics, threads and trimmings. Everything was in autumnal colours, a riot of colour to celebrate my favourite season.

That day will remain in my memory forever.

When I switched on the television at home I sat and watched as the announcement was made that our beautiful Queen had passed away. I

sat silently while tears streamed down my face. There were no words . . . I couldn't speak, and couldn't believe what I'd heard. It couldn't be true – she had always been there throughout my life.

My mind went back through the years.

The Silver Jubilee celebrations in my village.

A beaded amulet purse I made and sent to her as a present.

My uncle working for her at Wolferton Stud and then as Head Groom at Sandringham.

Watching from their window on a grey Christmas morning as my uncle wished her Happy Christmas as she went for a morning walk wearing her headscarf.

Times spent visiting the mares and foals at the stud.

I felt that although I had no direct contact with her that she was very much a part of my life and I thought she always would be.

I sat glued to the television and watched as events unfolded before me. I think like a lot of people I felt I had lost a family member; the country had lost its mother.

I sent a message of condolence to my uncle who had worked for her for many years.

I went into my shop the next day and removed my autumn window.

Replaced it with violet and black fabric with black ribbon. It was a very sombre atmosphere that day.

I visited Sandringham and left flowers and watched the funeral proceedings.

Still now I can't believe she's gone – a lady that was a huge part of my life has gone. Whenever they say "The Queen" I still think of my Queen.

There will never be another like her. Rest in peace Your Majesty.'

10

Paul and Emily
North Norfolk, England

'When Queen Elizabeth II passed away we were exploring Scotland in our campervan, a place we have come to love perhaps almost as much as our Queen did.

A matter of a few days before her passing we'd been in Edinburgh, learning about Holyrood and reading up on the Scottish traditions that The Queen so very much enjoyed.

As we were slightly off grid whilst camping we'd only picked up snippets of news but had come to, as many had, suspect she was unwell. We talked about how incredible a lady she was and how amazing a life she had lived, selflessly serving the country, often perhaps to the detriment to her own life and family.

On the 8 September we were making our way back down out of Scotland and into England. Hearing the news that we all dreaded brought a strange almost quiet stillness to the places we drove through, the country mourning together someone that so many of us grew up with. It feels an honour to have been in Scotland on those last few days of her life, to be in the landscapes she loved so much and found peace in, and then at her passing to be making our way back to North Norfolk, another place she was fond of with Sandringham Estate being that quiet space for her to escape to.

A few days later we visited Sandringham with family, sharing memories of our family's connections with the Royal Family and the estate, and saw the incredible outpouring of love and grief, with flowers stretching for miles upon miles.

Whilst we would never have labelled ourselves particularly as royalists, or found ourselves following closely the ins and outs of royal life, without a doubt our Queen held a special place in our hearts and her love for her country and her devotion to her people cannot ever be forgotten. She will be missed so very much.'

11

Samantha Thomas
Bristol, England

'The death of The Late Queen marked the end of an era, as she was the longest reigning monarch in British history. Her passing deeply affected the United Kingdom and the world with many reflecting on her leadership, resilience and role as a symbol of stability. It also triggered a period of national mourning and the shift in the monarchy when King Charles III ascended to the throne.

Her legacy as a global figure of diplomacy and services will continue to remain impactful.'

12

Jemima Agyemang
Blackpool, England

'I was in Ghana when I heard the news that Queen Elizabeth II had passed away. Even though I was far from the UK, the weight of the moment was undeniable. It was one of those events that made the world pause. I remember the murmurs in the streets, the discussions on the radio and the sense that history had shifted in an instant.

I definitely couldn't help but reflect on how much the world had changed during her reign. She had been a constant presence through wars, political shifts and cultural revolutions.

Her passing sparked different emotions – some felt deep respect for her lifelong service, while others reflected on the complex legacy of the British monarchy. For me, it was a moment to acknowledge the end of an era, to think about what she represented, and to wonder what the future would hold without her steady presence.

Even in Ghana, thousands of miles away, her passing was felt. It was such a reminder of how much history she had witnessed and shaped.'

13

Lauren
Ayrshire, Scotland

'I remember feeling saddened by the news of The Queen's passing. I also remember it being a huge topic of conversation at work and people having many mixed emotions about it. It felt like we had just celebrated the jubilee and then we were shocked with her passing.'

14

Jane Armstrong
Wrexham, Wales

'I was on my way to work when we first heard she had taken a turn for the worse. My heart broke – she has been a constant in my life and the lives of those I've lost. She holds a special place in my heart, a woman who put duty to us above all else. I even had a sticker of her in the back window of my car so it looked like I was driving her around. She was, and still remains, precious to me.'

15

Marie Hastings
Knoxville, Tennessee
United States of America

'When the public received the news of Her Majesty's passing, I was teaching a college class. Living in America, and surrounded by people who did not understand the love and admiration most Europeans had for The Queen, I did not expect much of an acknowledgement of the situation. However, throughout the day, dozens of my colleagues and students came by to see how I was doing. I'm glad God provided a silver lining to this extremely sad memory.'

16

Nicola Doyle
Glasgow, Scotland

'I had only ever seen Her Majesty once in person. I was taken aback by how elegant and nice she was. She launched a ship in my town and it was an honour and privilege to see her. I adored The Queen and was extremely upset when she passed away. I doubt that there will be another monarch like Her Majesty.'

17

Jeffrey Hyland
London, England

'Queen Elizabeth II had a close relationship with different parts of the Commonwealth, and the Crown Dependencies of Jersey, Guernsey and the Isle of Man were no exception, so it was somehow fitting that I was visiting Guernsey when The Queen died.

I first heard the radio reports suggesting that The Queen's passing was imminent coming from a loudspeaker in a small gift shop and raced back to my hotel to watch the TV news coverage. When the inevitable news came, I felt a huge wave of emotion and the tears welled up. As a dedicated royal watcher and writer for all of my life, I felt that I had a special relationship with The Queen, even though I had never met her in person. My phone started ringing as first my younger brother and then other friends and family rang or messaged me to say they had heard the news and the first person they had thought of was me, as they knew that I had dedicated so much time to studying the Royal Family over the years.

The next day, we gathered in the small harbour at St Peter Port, the capital of Guernsey, to hear the Noonday Gun fired from the ramparts of the historic Castle Cornet. Traditionally fired to mark 12 o'clock each day, this time the cannon was fired a sombre 96 times – once for each year of The Late Queen's life. Hundreds of people came out from the town down to the quayside to pay their own respects to a special person as the cannon fire echoed across the harbour.

When I returned to London, my attention turned to my work for the Commonwealth Parliamentary Association (CPA). As Head of the Commonwealth, Queen Elizabeth II had supported the organisation since her accession to the throne in 1952 and later became its Patron in 1989. The Late Queen had visited almost every Parliament in the Commonwealth during her long reign.

Tributes and acts of remembrance poured in from many of the 180 national, state, provincial and territorial Parliaments across the Commonwealth's 56 countries as well as thousands of Parliamentarians who were also paying their own personal tributes to The Late Queen. These ranged from the Parliament of Sri Lanka holding two minutes' silence in the Chamber at the start of their parliamentary session; to Parliament House in Australia, where a 96-gun salute and flags flown at half-mast marked the death of The Queen of Australia; to the Parliament of Bermuda where a special sitting was held for Members to pay their own tributes with the President of the Senate, Hon. Joan E. Dillas-Wright recalling being a school pupil waving her flag enthusiastically to greet the young Queen on her first visit to Bermuda in 1953; and in the UK Parliament, many MPs and Peers formally pronounced their renewed parliamentary oaths to swear allegiance to the new King, followed by tributes in both parliamentary Houses to the late monarch.'

18

Joanne Greensill
Wolverhampton, England

'I remember being at work with BBC News on the television in the staff lounge at work. The presenters were reporting on The Queen's health, and as the day went on, we started noticing subtle but telling changes – the news reporters started switching to black dresses, black ties and black suits. It made us all think that the time had come, and we were right.

When I got home, I spoke with my husband and the kids, and I called my mom. It felt like a very sad day, a profound loss for the country. The Queen had been a constant presence throughout my life. I truly admired her and saw her as a remarkable leader, both for her family and the nation.'

19

Gary Simmons
Sheffield, England

'I first became aware of the events of the day during an afternoon conversation with a colleague who shared the news that Buckingham Palace had shared significant concerns regarding the health of The Queen. We discussed how momentous a day this could be. There was a quiet atmosphere in school as the news was being digested.

I was driving home listening to the BBC Radio 4 coverage when Mishal Husain at 6.30 p.m. announced that Buckingham Palace had just confirmed that The Queen had died. It was one of those moments that you will always recall where you were when you heard the news.

I said a little prayer of thanks and reflected on the remarkable life of service that The Queen had given her nation. She was the bridge from the past to the present, spanning multiple generations. Winston Churchill was her first prime minister with Britain still rebuilding, still under rationing following the Second World War, and ended as a 96-year-old Queen in the digital age. Yet she never felt out of date or out of touch.

Her life of duty was appreciated by all – monarchists like myself and non-monarchists alike.

Her death, as all deaths do (to some extent), remind us of those we have lost and this was no different. I reflected on how my mum (who died in the January of 2022) would have felt, having hugely admired Queen Elizabeth II.

Once home I watched the BBC coverage of her death and was impressed by the dignity shown by those broadcasting and impressed by the smooth transition.

I remember listening to Prime Minister Liz Truss with interest and being shocked at her announcement of King Charles III, not just because there had been speculation that he might take a different name given the history of the previous two kings named Charles, but primarily because it was the first time in my lifetime that I had heard anything other than Queen being uttered.

Former Prime Minister Boris Johnson released a statement that included something like the second great Elizabethan age. I haven't often agreed with him but I would concur that Elizabeth II was a great monarch. I reflected that Queen Elizabeth II had lived a remarkable life, seen enormous change in her country and in the world and yet retained her commitment to duty to the very end. A fitting epitaph.

In the words of the new King, 'God speed Her Majesty.'

20

Andrew McDaniel
Caithness, Scotland

'She was Queen of Scots and we all knew of her great love of the Highlands.

When I heard the news and thought that her final days were spent along the River Dee, I smiled. Especially because Royal Deeside brought her such peace and tranquillity. May she rest in peace after doing so much for so many.'

21

Martyn Dore
London, England

'A few days after her death, I flew home from Sitges, the captain sent his condolences to all British passengers aboard the flight, but it still didn't seem real.

I think it was only real when I got home, and started to see announcements on posters from businesses sending their condolences, that it finally sank in that she was gone.'

22

Gail Popock
Cwmbran, Wales

'When I received the official notification of her passing I was in the middle of a badminton game with my husband at our local gym and I couldn't believe my eyes at what popped up on my smart watch! I had to take a moment aside and have a little cry. Her funeral was held on the day we flew to Florida on our honeymoon!'

23

James
Dorset, England

'It was such a shock to me. I just thought she would live forever.'

24

Amy B. Garratt
North Lincolnshire, England

'I was heading to Doncaster for a spoken word evening. My husband and I met up with some friends beforehand at a local art gallery. We were roaming around looking at all the various art, and the radio was playing on one of the floors. Suddenly a murmured conversation seemed to pass from person to person in the room, and I was trying to figure out what was going on. Then this voice suddenly and very loudly announced "SHE'S DEAD!" (Quite surreal!) Everyone just went quiet, not really knowing how to respond to the outburst, and we established more quietly that The Queen was the one they were talking about. We said at the time, "It's strange to know when you're living through a historical event and a momentous moment during everyday life." We wandered over to the spoken word event talking about the only monarch any of us had known.'

25

Andrew Lannerd
London, England

'Sitting in a small café in Ballater, Scotland, I was enjoying what I thought was a routine coffee during a busy day. Instead that moment started a sequence of events that I knew would eventually come, yet had been dreading since the age of 13.

"Such a shame about The Queen's health," were the words I overheard from behind the café counter. I immediately pulled up the BBC app to view the breaking news from Balmoral, which was only down the road from where I was sitting. Looking at the news it became clear that The Queen's health was in decline and that members of the Royal Family were already *en route* to Royal Deeside.

We all need someone to look up too, especially as children. My own story with The Queen started when I was 13 years old. Growing up in small town USA, I often felt out of place, and the world around me seemed very insular. The many centuries and layers of British history became the focus my young and inquisitive soul needed. Quite frankly Queen Elizabeth II and the British monarchy sparked my lifelong interest in the wider world.

Such was my interest and enthusiasm that I uprooted my American life years ago, and I now fully reside in the United Kingdom.

In recent years I was fortunate to see The Queen in person on many occasions, and even had the joy and memorable experience of sharing a laugh or two with her. Some of these times I was in the crowd alongside one of her many royal visits. During later occasions I was lucky enough to be an invited guest. Regardless of the type of event, The Queen's presence and marvellous smile never failed to radiate those around her with joy.

I cherish those memories. Despite the ancient exalted position that she held, I witnessed her gracious kindness, humility and warmth as a human being.

I always thought of The Queen as living history representing the best of Britain. A human connection to 1000 years of history – yet despite the majesty of her calling, her smile, wave and humanity are the things many of us will remember the most.

I will never forget the time I heard The Queen say, "Don't you all look lovely," to friends of mine, or the occasion at the age of 93 when I watched her go out of her way to make a child feel special when she was probably exhausted at the end of a busy day of engagements.

Leaving the café on 8 September 2022, I walked the short distance over to Ballater's Glenmuick Church, a place of worship that The Queen knew well. I opened the door to a quiet, empty sanctuary and slowly walked up to the altar where I said a prayer that God would be with her and us during the days ahead. At that moment I could already feel the weight of the news that was still to come.

Later that day like so many millions around the world, I was glued to the television coverage. I can still see the moment on the BBC when they showed a half-mast Union Flag flying over Buckingham Palace. Then the next second the sorrowful announcement was made.

I immediately felt a painful sense of loss. My own chapter with Queen Elizabeth II had ended.

It still feels fortuitous that even though my interest in The Queen began thousands of miles from Britain, I was near Balmoral, The Queen's much-loved Scottish retreat when her long reign drew to its close.

Three days later in Edinburgh I witnessed the proclamation of King Charles III, and that same afternoon the procession of The Queen's coffin up the Royal Mile from the Palace of Holyroodhouse to St Giles' Cathedral, where The Queen would Lie at Rest.

These events had never happened in my lifetime, nor the lifetimes of so many gathered.

As The Queen's cortège came closer, a dramatic silence descended over the enormous crowd. I could hear the guards' footsteps getting louder and louder as they took their respectful Slow March up the hill closer towards me.

I deeply bowed my head the moment her coffin, draped in the Royal Standard, passed by.

That evening I laid flowers at the gates of the Palace of Holyroodhouse. My handwritten card encapsulated what I was feeling then, and still feel today: "Words cannot express our thanks for what you meant to us. Thank you, ma'am.'"

26

Georgina Wright
Blackpool, England

'It started out like a normalish day. Email in the morning, approximately 300 cups of tea and working away on various projects.

Then the BBC News app on my phone pinged.

I saw the headline stating that Queen Elizabeth II was unwell and members of the Royal Family were being summoned to her location, and I just knew.

I spent the rest of the day in an anxious bubble, not sure how to feel. I knew that any updates could mean making changes to work. As we're local authority-based I already had some idea of media blackouts so I quickly rang some colleagues and asked for their advice on what/if I needed to do anything. (This was still before extended royals had been called up to Balmoral.)

The rest of the day passed in a blur of emails, team chats, news stalking before I finally went home. My husband and I had the BBC News on, just watching quietly, waiting.

I remember writing a status on my Facebook – "God save our Queen."

The dog did her usual post-dinner dance at the back door. I let her out for a minute, came back in and saw the half-mast flag on the screen.

I asked my husband, "Why is the flag like that, what did I miss?" It had changed in the seconds I was coming back into the room, though I'd barely finished my question before the black BBC screen came on and Huw Edwards then told the UK and beyond about the passing of our late Queen.

Even though there had been a whole day of news updates – and the fact that Buckingham Palace rarely share such details was the biggest soft approach to this news – it still washed over me like a wave of sadness.

The only thing in society that had stayed the same, unchanged throughout my entire life, was now changing; it felt quiet, empty for a moment.

I silently sent a prayer up, thanking her for her service before snuggling up to my husband and watching the news continue to break.'

27

Wendy Robey
Devon, England

'My daughter and I were on our way home from IKEA after stocking up on essentials for her first year at university. We'd been back in the car for only a few minutes when the news came over the radio. We both looked at each other and started talking about her life and what a change it would be not having her on the throne.'

28

Tim Jukes
Dudley, West Midlands
England

'I was sitting at home with my wife watching the BBC when the news of her passing was announced. Everyone was wearing black a few minutes before, which was a giveaway to what was about to unfold.

I'm not a royalist but I did have a sense of loss as The Queen was a constant in my lifetime and very much respected.'

29

Jade Nomura
Taunton, England

'I was deeply saddened with the news of Her Majesty's passing. I remember just watching the news and thinking this was the end of an era. It's still sad to this day. She was such an inspirational woman; I think she was a brilliant Queen and icon for our country.'

30

Steph Barker
East Yorkshire, England

'I was away in Iceland at the time with my husband when we heard the news of the Queen's passing. I remember Icelanders overhearing our accents, realising we were British and then offering us their sympathies. I've always been fairly indifferent to the Royal Family, but I've always had a huge amount of respect for Queen Elizabeth II and was deeply saddened by her passing. Her reign was incredible – she was witness to, and accomplished so much throughout her life. She was a woman you couldn't help but love and respect, and we lost a remarkable woman and monarch that day.'

31

Emma Nankervis
Brighton, England

'I was working a long day shift at work so hadn't been looking at my phone or listened to the radio due to how busy we were. When I got back to the office, I noticed I had 5 missed calls from my mum so called her to see what had happened. She told me the news that The Queen had died and we both cried over the phone.'

32

Charlie Morris
Cornwall, England

'I remember being at home and watching television as normal when the breaking news came on. I don't usually follow the Royal Family but it was so emotional to watch. I even shed a tear! But at the same time, it was amazing to be reminded of her legacy and the brilliant things she has done over the many years. She really was a remarkable monarch!

My brother-in-law was also one of the servicemen involved in the funeral ceremony, which was a beautiful send-off. It was heartwarming to see so many come together for it.'

33

Elle Padilla
Bracknell, England

'I was celebrating a birthday when The Queen died. It was so surreal, almost unsettling, to be toasting another year of life while the world mourned the end of one. There was an unspoken weight to the air, as if time had briefly paused to acknowledge the quiet symmetry of it all; where one life was honoured in remembrance, another was being celebrated in its continuation. The contrast was stark yet profound, a reminder of life's relentless rhythm, its beginnings and endings forever entwined. I felt very mournful that day.'

34

Robert Bailey
Sydney, Australia

'As an Aussie, The Queen was really part of our DNA you might say. She was always there, and we all carried around a little of her in our pockets.

I heard The Queen died when turning on the radio. The morning instantly changed for me, and so much had happened while I was sleeping.

The day went on as usual in Sydney – yet did it? It felt different I must say. Good on ya, Ma'am – you did us proud!'

35

Diane Clarke
Norfolk, England

'I started following the story on the news when they said she was unwell. I had booked a session at the gym, so off I went. Myself and my personal trainer were watching the news. By the time I walked home and checked again, the news reporters were wearing black so I knew she had most probably gone. My youngest son (a teenager) and I watched the news together as they announced her passing. Felt surreal, as she had been The Queen for so long during my life. I had never known another.'

36

Alex Clement
London, England

'I work at a theatre in the West End as part of the front-of-house team. During the pre-performance briefing we were aware of The Queen's decline and we were told that the shows were to go on until told otherwise. During the second show of the day (which ironically was about six British queens), we were told that The Queen had died over our internal radios. We had to keep a normal face during the show, and then an hour later at the end of the musical, everyone began looking at their phones. Many of them started to cry and we had to try and keep them calm as they left. It was a unique situation to say the least. For about a week after her death, before each show, the stage manager would come on to the stage and mention The Queen's passing then the audience would sing the National Anthem.'

37

Peter Steer
Douglas, Isle of Man

'I couldn't imagine a life without The Queen on the throne. How would her death impact the world, the Commonwealth, the British Isles, me? It was unthinkable that she had gone. I suppose having lost most of my family The Queen had become even more so the representative of authority and wisdom.'

38

Danielle Amor
Berkshire, England

'I am a huge supporter of the Royal Family and being a government officer we had to go into State Mourning, so had a lot of work to do cancelling public engagements and meetings.

I was on a training course for work (on Zoom) and I saw the BBC newsflash that the senior royals were rushing to The Queen's side, so I quickly put the BBC News on television as well. I texted my husband and said should we go to London as I thought she was going to pass away, to which he said I was mad – of course she wasn't going to! As soon as the course finished, I sat in front of the television for the rest of the evening and was so upset when the screens went black to announce that she had passed away. Suffice to say I don't remember much of the online course because my mind was distracted with the news!'

39

Fiona
Downpatrick, Northern Ireland

'So many times during her reign The Queen talked about peace, friendship amongst neighbours, and goodwill towards all.

How lucky we were to have that kind of leadership – demonstrated fully when she worshipped in both the Catholic and the Protestant churches that day in 2012 in Enniskillen.

Her doing that made a difference, and I wish I could have told her just how much.

Her death was memorable to me for many reasons, but it's rather hard to forget how a small group of Protestants and Catholics gathered together the Sunday after to give thanks for the life she led.'

40

Adam Boulton
Somerset, England

'I remember being told about it as I walked to the pub and thought how strange it would be to have money with a different face on it from now on.'

41

Toni Broom
Leeds, England

'The day before The Queen passed I had to take my beloved dog, Harry, to be put to sleep. I was already mourning but The Queen's death hit me harder. All the constants I had expected and knew were gone in less than 24 hours. I struggled to see the world as anything more than death. It's taken me a few months to see the joy in life again but I'm getting there.'

42

Lauren Sparks
Liverpool, England

'I was parking my car and the news came on unexpectedly, and I burst into tears.'

43

Barry Stewart
London, England

'I was in Eastbourne to see *The Mirror Crack'd* on stage. I had seen Joe McFadden jogging along the seafront and said, "What ho! Looking forward to seeing you on the stage tonight. I shall be in the pub next door." Later that day I received word that the show was cancelled and the pub was closed! They dimmed the lights on the theatre as well, which felt rather sorrowful. I watched the news all night in my hotel room.'

44

Natasha Bayliss
Hampshire, England

'I was at work and we saw the Royal Family travelling and everybody commented that she had probably already died.

We were stunned how The Queen had been working just days before when we find it a slog to make it to Friday some weeks! We all felt that it was the end of an era.'

45

Andrew David Stedman
Northumberland, England

'All the people outside the palace, damp and squelching, slightly stunned and confused, lights of London smudging strange reflections in all the puddles. A dark and cloudy night.'

46

Ken Stone
Ipswich, England

'I really became aware of Her Majesty when watching the 1960 wedding of her sister, Princess Margaret, on a black-and-white television. The following year, The Queen undertook a tour of India and Pakistan and we were encouraged, at school, to follow this. Then on 22 July 1961 Her Majesty came to my home town and I saw her in person for the first time. My now late mother and grandmother had collected magazines and books on the Royal Family over many years, so I have continued the interest for now over 60 years, particularly during the 1970s making scrapbooks of newspaper cuttings of the extended Royal Family. Photography then became a hobby and I have now some 5000+ transparencies from pre-digital times to sort through, many of The Late Queen.

Living relatively near to Sandringham and not too far from London, gave me the opportunity to photograph Her Majesty on hundreds of occasions with the privilege, especially in the Sandringham area, to share a few words in person with her. I particularly recall one occasion during a visit to Cambridge when The Queen and the Duke of Edinburgh had arrived by train to visit the university. There was a very much environmentally friendly focus on this visit that necessitated the royal visitors travelling on a 'green' bus to the first location. As Her Majesty approached me and exchanged a few words, my last comment was, "I hope you enjoy your bus trip, Your Majesty." Instead of a positive verbal response, she just shrugged her shoulders and grinned! *Duty as always*, I thought to myself, and something which summed up the commitment made on her 21st birthday – "I declare before you all that my whole life, whether it be long or short, shall be devoted to your service and the service of our great imperial family to which we all belong." – and it was.

On the day of The Queen's death, my wife and I were staying at Stansted Airport in readiness for an early flight to Copenhagen. We had followed news of Her Majesty's declining health during the day but when the news broke on television, the whole lounge where we were fell silent – almost in disbelief that what we knew would happen one day, had happened. I reflected on the news and felt grateful to have been born in Her Majesty's Accession year and to have been an Elizabethan my whole life.

Upon my return from Copenhagen, I went to London and witnessed the return of Her late Majesty from Hyde Park Corner. The next day I returned to watch the magnificent procession from Buckingham Palace to Westminster Hall from a vantage point in Whitehall, opposite, very appropriately, the Memorial to the Women of World War II in Whitehall. Later in the week I went to Sandringham for the visit of the new Prince and Princess of Wales to view the mass of floral tributes.'

47

Alyssa Lamanna
Isle of Wight

'I was at bingo the day The Queen died. While there I saw the news that the Royal Family had been called to her side. By the time we got home in the evening she had passed. I remember just feeling so shell-shocked that our Queen had died. It just felt like she would live forever and I felt so sad for myself, for the Royal Family and for the country.

But watching the Coronation of King Charles III renewed a sense of hope I felt had been lost.'

48

Heather Crosskey
Wallington, England

'I like to watch the news whilst eating my lunch – it's become a habit. 8 September 2022 was no different, except I'd switched the television on before I'd finished cooking my meal. What met me was a very sombre newsreader and the headline that doctors were concerned about The Queen's health and she was "resting comfortably" at Balmoral. *Well*, I thought, *she's had a busy few days getting rid of one prime minister and appointing a replacement, I'm not surprised she needs to rest*. I got frustrated with all the talk about her being old and frail and comments about the bruise on her hand. *Yes*, I thought, *she is old, but that bruise could be as a result of knocking against a chair or table as she walked past.* (As a retired nurse I know how fragile aging skin can be).

I carried on watching throughout the afternoon. Much of the information being given was just repeated in between little video snippets of Her Majesty attending the various engagements she was able to do in the months beforehand and there was talk about the Platinum Jubilee celebrations earlier in the year.

I started to feel a bit more concerned when they talked about "the family gathering" at her bedside and watched as Princes Andrew, Edward and William arrived, along with Sophie, Countess of Wessex. At that moment I had the feeling that our beloved Queen had already passed away and only the Princess Royal, the Prince of Wales and the Duchess of Cornwall had been present at her bedside.

I'm not ashamed to say I cried when Huw Edwards made the announcement that "Her Majesty Queen Elizabeth II has died peacefully at Balmoral." It was the end of an era. It seemed odd to grieve the death of someone I didn't know, had never met, but who had been there my whole life.

I think I watched every second of every programme about The Queen from that day until her funeral. I was sad, but so immensely proud of our Royal Family. We had lost our Queen, but they had lost a beloved mother, grandmother and great-grandmother. We are able to grieve in private when we lose a loved one, but they're not able to – King Charles was back in London within 24 hours of becoming King, met the crowds outside Buckingham Palace, recorded his address to the nation and made a brilliant job of it. The week that followed saw him travel the length and breadth of his kingdom as he attended the various proclamation ceremonies and official events. Other members of the Royal Family met with people outside Balmoral as they viewed the tributes and flowers folks had left.

I'm a monarchist and proud to declare it. No, our Royal Family isn't perfect, they have problems like any family, but that's what makes them seem more "normal" to the rest of us.'

49

Emma Mortimer
Taunton, England

'I was at work and my manager called me into the office. When she broke the news, I cried. She then went to put on the kettle. Once tea was made, we chatted about The Queen's amazing life and how wonderful she was. Everything's better with a cup of tea. We then had to break the news to the rest of the staff.'

50

Jacqueline Surridge
Lancaster, England

'On the day that The Queen died we as a family were having a day out in sunny Morecambe. We went to the festival market to do a bit of shopping and my son Thomas had seen on the BBC News app that The Queen's family were gathering at Balmoral. We all began to speculate that it must very serious for that to happen.

There were a lot of people around but it didn't seem to be a topic of conversation – mind you, Rishi Sunak had been flown in by helicopter and was posing with the Eric Morecambe statue. Anyway we went about our day and returned home. We put the television on and more or less straight away a very grave Huw Edwards began to tell us she had died. I immediately rang my daughter-in-law Beth to tell her and Thomas the sad news. They thought that I was joking and couldn't believe that it had happened. That evening we all went out for a meal in Caton; the mood wasn't sombre or anything, but it had a really strange feel about it. It felt really odd to think that she was no longer our reigning monarch and that Charles was King.'

51

Stewart Davies
Argyll, Scotland

'I don't even have or use Twitter but somehow, and I can't even remember why or how, I read a tweet on my iPad – "Queen dead, 17 seconds ago". Rushed downstairs to say, "Oh my god, I think The Queen has died."

Soon as I opened my mouth the breaking news came on the television and then there was just a hush. Like a tangible pit in everyone's stomach. Horrible news. You're not allowed to love The Queen any more without getting cancelled, but I don't care, we all adored her. The nation's grandmother and I was gutted for the whole week, watching and recording every bit of the coverage.

Particularly poignant and surreal night watching her coffin on TV arriving in London and just seeing endless crowds of people lining the streets, traffic stopped, people abandoning their cars on the motorway to stop and watch. Surreal moments I'll probably never forget or experience again.'

52

Michelle Nicholls
Claydon, England

'On the day it was on the news that her family had been called to Balmoral, I was at the beautician's having a treatment.

My stylist and I spoke about the situation and I said, "She's gone." Then literally a few minutes later the news broke – The Queen had died. We both cried.

It was so surreal. I never met The Queen, but she'd always been a part of my life. It was like we'd lost a member of the family and we knew things would never be the same again!'

53

Jemma Hole
South Wales

'I was working but I had the news on all day on my laptop to hear the updates. I felt weirdly anxious about it all (considering this was a person I'd never met) but it felt so profound. My partner and I drove to his parents for the evening, and we were all chatting about The Queen, and I was saying how I thought she had already passed – it seemed the news people knew as they were in black ties and so on. The news was on in the background, and then we all stopped talking when we saw the banner on the news and said, "Oh my god, she's gone." The golf tournament we were due to attend next day was cancelled out of respect and I spent the whole next day watching the news.'

54

Kelly Holland
Hampshire, England

'I was at work when The Queen died. I remember someone seeing the news feed pop up on their computer and it soon spread round the office. It was quite a poignant moment as our manager decided that we should do one minute's silence as a mark of respect. It was odd knowing she was no longer with us and quite emotional seeing all the tributes on the television. My grandad actually passed away on the day of The Queen's funeral. It was playing on the TV at his nursing home and he took his last breath as The Queen's coffin was carried out of Westminster Abbey.'

55

Gareth Hamilton
Riga, Latvia

'I was teaching at a university in Riga. I had to be there all afternoon before teaching three hour-and-a-half lectures from half past four until half past nine. I checked the news all afternoon before having to go and teach. It was a strange feeling wondering what might happen while talking to students and not being able to follow the events as they happened. This was the second day of the teaching semester and I was with Master's students. I did mention the events happening and we noted it was very rare and unusual, of course. Then I said I would actually check the news and that was how we all found out. There was shock. It was quite an international crowd with students from Latvia, of course, but with others from China, the USA, Germany, India and other places. With migration and studying abroad, a person in the class had been in the UK for their Bachelor's studies and her husband is British so she was even more astounded due to her connection with the country. I then had to wait to find out more – that is, to get home to catch up with the news. I got a taxi and it felt very strange and melancholy sitting there in the back, waiting to get more information. As a footnote, on the day of the funeral I was invited to the British Embassy to watch it on television, which was a nice bit of closure.'

56

Sean Weston
Derby, England

'The day The Queen died, I was on a cruise ship, touring the Norwegian Fjords. I'd seen on the news that day that The Queen was unwell and that the Royal Family were travelling to Balmoral. I just stepped into the bathroom to get showered and ready for dinner when an announcement came on in the cabin. This is quite unusual and the captain announced that The Queen had died. I just stood at the sink in shock. My husband shouted through the door to ask if I was OK. I managed to say yes, I was and realised I was crying. I think it was the fact that she had dedicated herself to her role and it just hit me that she was gone. The ship was very subdued that night.'

57

Russell Deathridge
London, England

'The mood in London felt strange (or was it just me?), knowing that we were part of history unfolding. It reminded me of the famous line "The King's life is drawing peacefully to its close," released by Buckingham Palace as George V neared his death. I remember looking across both sides of the city as I walked across London Bridge thinking, *Another notable day this old city has seen.*'

58

Fiona Brown
West Midlands, England

'Colleagues and I finished our day of teaching and were speaking about the news of her illness. We finished at 3.00 p.m. so this would have been just after and I had a feeling she'd already gone when we were chatting. It turns out she died at 3.10 p.m.'

59

Saif Mokeef
Derby, England

'I've never considered myself a royalist but when my grandma told me The Queen had died I felt my breathing change.

I don't know why but I never felt that way before. We were decorating a bedroom so didn't know she'd died until two hours later when we came down for a tea break.

I had to sit down in the kitchen for a while and the whole family just sat in silence for a few minutes, not necessarily out of respect but because our eyes were tearing up and we didn't know what to say.

My grandma told me that when she was a child, Queen Elizabeth was also Queen of Pakistan so she felt a connection and sadness in regard to her in that way.

I was fixated on the news after that until the day of the State Funeral. I don't think I remember ever feeling that sad prior to her death even though I'm 18. I had the same emotions as when my great-grandma died.

Now I do consider myself a royalist because I consider Elizabeth II to be the greatest monarch in English history.'

60

Debbie Ioanna
Bradford, England

'I remember it was a Thursday. First thing that morning, I'd had a wisdom tooth removed, so I was already feeling delicate (and sulking if I'm honest!) due to the pain.

I can't remember what time, but it was reported about The Queen's condition and senior royals were on their way to be by The Queen's side. So, from then on, we had BBC on constant. Huw Edwards was on all day and it was repetitive but none of us at home wanted to admit we knew what was coming, we were just waiting for it.

I was messaging a friend at the time and at 5.30 p.m. she noticed that Huw had changed into a black suit. Nothing had been announced but that was quite indicative of what was shortly coming.

I didn't want to go to the bathroom, didn't want to get a drink, didn't want to leave the television, waiting for the now respectfully dressed Huw Edwards to announce what we already knew was coming.

But, of course, I have a small child so sitting watching television undisturbed doesn't work. She wanted juice so, in roadrunner fashion, I darted to the kitchen ... only to hear, "Queen Elizabeth II has died ..."

All blooming day in front of the television and Huw waits for me to disappear for a split second!

So it was announced – obviously an upset house (apart from my child who was unaware of the significance). But my job, which I'd only been in a few months, as admin and social media for a parish council, meant I was suddenly on overtime amending website announcements, social media posts, everything so rules and protocol were followed. It was a rush, it was so sad, and none of it helped the excruciating pain in my mouth.'

61

Hassan Jain
Delhi, India

'In the busy streets of Delhi there isn't much that grabs the attention of people. So many people just moving about and getting on with their work and daily lives.

That was different when word went around the streets that Queen Elizabeth had died.

She was the Head of the Commonwealth so we all felt we knew her. A true leader of the world was mourned that day.'

62

Anonymous
Student from Newcastle, England

'I remember where I was.

It was a Thursday and the date was my monthly swing dance social that evening.

I remember seeing the news and crying. Mostly out of frustration because I had lived through more than enough historic events in the prior four years than many people would live through in their lifetime.

It was strange, later on in the day, walking down an empty Northumberland Street, with every electronic billboard either black or displaying her face. That will forever be an image I will never forget.'

63

Frederick Hervey-Bathurst
Hampshire, England

'Benjamin Britten is often referred to as England's greatest composer since Purcell and in 1962, he wrote a new arrangement of the National Anthem "God Save The Queen", for the Leeds Festival. The decade produced some of Britten's greatest works and coincided with a phoenix-like revival of the British arts scene from the wreckage of war, collapse of empire and national bankruptcy. Britten's great *War Requiem* was composed for the consecration of the new cathedral at Coventry (replete with sculpture by Epstein, tapestry by Sutherland and glass by Piper) whilst his National Anthem became the soundtrack to the openings of the Queen Elizabeth Hall at the boldly brutalist Southbank Centre and at Snape Maltings, home of the Aldeburgh Festival, both in the presence of The Queen. In this sense, these two compositions lie at the very centre of the artistic resurgence that marked the early years of The Queen's reign. Britten's National Anthem reimagines the rather stolid tune into a work with almost symphonic progression despite its short length. The choir alone begins the first verse with a sense of hushed and breathless awe before the orchestra joins, climbs to a new key and leads the choir in a triumphant and declamatory rendition of the second. Allegedly, it was The Queen's favourite arrangement.

On the 8 September 2022, I was working in the City of London as the BBC began to send its news alerts: first, generic and ambiguous statements on The Queen's health and then increasingly frantic as the world began to track the specific movements of the other royals to Scotland. By the end of the working day, Nicholas Witchell seemed to be in a state of mawkish delirium and speculation was rife about the contents of those whispered communications in the House of Commons. I had worked at a magazine a few years previously when similar (and then erroneous) rumours had abounded that the Duke of Edinburgh had died, inciting similarly fruitless speculation. So, only being a tepid royalist myself, I logged off, put away my phone and set my mind on that evening's rehearsal for the Last Night of the Proms. This grand event, held annually in the Royal Albert Hall and broadcast to many millions on radio and television around the world, marks the end of the world's largest classical music festival with an evening of Britain's greatest patriotic music. As a member of the BBC Symphony Chorus, I boarded the new Elizabeth Line in the direction of the BBC studios in Maida Vale where the penultimate rehearsal was due that evening before the concert itself in just two days' time.

The Maida Vale Studios complex was built in the 1930s inside the shell of an old Edwardian skating rink and there is still a charming sense of the 'make-do-and mend' about it; it is certainly a long way from grand. Despite this, its cultural history is unrivalled. Not only did the studios host BBC News during the war (Broadcasting House's central location was too susceptible to the Blitz) but almost every great musician over the last 90 years has performed or recorded there.

When I arrived at Studio 1, the room was filled with the usual pre-rehearsal noise of the orchestra tuning up, the scraping of chairs on the staging and animated conversation. The start time of the rehearsal came and went as we sat in position. At the other end of the room, Dalia Stasevska, our conductor who had flown in from Helsinki, talked earnestly with the BBC officials who came and went through the swing doors clutching clipboards. After a while, she walked forward to the podium and announced the news of The Queen's death and we held a minute's silence.

Though I am not a republican, I have never been a particularly fervent monarchist. I find the history of the institution fascinating but the sycophancy and melodrama of so-called royal commentators has always grated with me as has the prominence of the Royal Family in our national media. In short, I am largely indifferent to the twists and turns of the national soap opera that is the personal lives of our Royal Family, so I was surprised by the effect that the news of The Queen's death had on me.

During the minute's silence, I looked down on to my lap where I held the choral score of Britten's *National Anthem*, this piece that so encapsulated the cultural renaissance of the "new Elizabethan Age". It was Elizabeth herself and the youthfulness of her early reign that in some ways transmuted the stiff culture of Britain's pre-war artistic establishment into the bold and dynamic ecosystem of the 1950s and 60s. This daring spirit of a new aesthetic epoch with The Queen as its figurehead was epitomised by Britten's *War Requiem* and *National Anthem*. Yet Britten had died over 20 years before I was born and over 45 years before now. He was history and yet The Queen had lived on, her presence and surety unchanging, a constant in the background of British everyday life. Now that was over and it was only then, with her leaving, that I began to realise the depth to which her simple presence had permeated our own age.

Dalia explained that she did not know whether our performance would be cancelled or not, but that, now we were all here, we might as well rehearse. As we sang Britten's marvellously sensitive harmonies, I reflected that this was probably one of the few arrangements of our National Anthem ever written specifically to the words "God save the *Queen*" and it was likely that I would never have cause to sing those words again in my lifetime.'

64

Robynne Williams
Wales

'I was with my son at home when the news announcement came on.

I remember growing up with my great-grandmother, who loved The Queen. I watched her Christmas Day speech every year.'

65

Hayley Anderton
Preston, England

'The day The Queen died it felt like the end of an era. Growing up I thought she was the most powerful person in the world, and I guess I always carried a high opinion of her, even as I learned more about the monarchy. I felt sad, and I knew the news would have a big impact on the older generation in my family. I don't think the Royal Family will ever be quite the same in our changing society, and it feels like she left a hole in the monarchy that none of the royals can repair.'

66

Nicola Smith
Bristol, England

'I remember the day our Queen passed away like it was yesterday. I was on a shift at my job as an administrator. The day went well. No hiccups and thankfully, all in all, a good day, or so we thought. I remember that morning the news saying that our beloved Queen had been taken poorly and the Royal Family were going to see her. Knowing this, we all had the BBC News on our laptops awaiting further updates. Later in the day, the presenter of the news changed into a black suit – we all then took a gasp.

We knew in our hearts our beloved Queen had passed. As I finished work, I rushed home to continue watching the news. At 6.30 p.m. it was announced that our Queen had indeed passed away. I cried my eyes out, for I had always known The Queen, how hardworking she was, and she was a hero to many.

That night, I had a gin and tonic and raised a toast to Her Majesty.'

67

Emma
Student in Scotland

'I was, quite literally, packing to move into student accommodation for university when my mum called me through to her bedroom. It was known that The Queen was unwell but "comfortable and at rest".

Immediately, the whole family thought "OK, The Queen has died," as it seemed obvious by the statement made.

Then it was announced that she had died, and it was both . . . like, this somewhat inconsequential thing – it's not like we knew her personally – yet it was also an insanely huge thing. She'd been in power for so long, and it was kind of overwhelming.

Since my university wasn't far from Balmoral, emails started flooding in, as we were still to move in two days later, but now heard that there'd be a police contingent staying in our student accommodation the night before to travel with The Queen [to Edinburgh], which felt kind of crazy.'

68

Davina Woodhouse
North Yorkshire, England

'I was at home on maternity leave with my then two-month-old baby. I remember searching news sites for information and updates. I was desperate to go to London to pay my respects but with a 2-month-old and a toddler it was impossible.'

69

Sally
East Midlands, England

'I always felt The Queen was incredibly hardworking and she was always there. I remember feeling shocked and surprised by her death.'

70

Al Jones
Shropshire, England

'I was at home with my children and saw the news of The Queen's passing and thought how very sad it was that she had died, as I would anyone.

A family had lost a mother, a grandmother, and that's always sad to me regardless of who it may be.

Still feels very odd not hearing *God Save the Queen* during the National Anthem but life goes on.'

71

Stephanie Wadsworth
Nottingham, England

'The day I found out about The Late Queen dying I was at home with my husband and my two boys were upstairs. Randomly we had just helped our neighbours move a dining table they had bought from us, so we were back and forth from each other's houses. We knew something was happening as there had been a lot of speculation throughout the day, so each house had the television on.

Then the news reporter came on and told the nation the news. I remember crying and being typically British decided to make a cup of tea. Meanwhile my neighbour went over to open a bottle of bubbles to celebrate her life. We just sat around talking about The Queen and what would now happen for the country. It was all very sad but nice at the same time.'

72

Matthew
West Midlands, England

'I was at home watching the news. It was like losing a family member.'

73

Stuart Cummins
Hove, England

'I remember putting the news on and immediately knowing that The Queen must have passed away. It was a surreal moment, as she had been on the throne for my whole life. Whilst it wasn't a complete surprise, it still felt like the end of an era. I have many fond memories of times wrapped up in celebrations related to The Queen. My two young daughters were absolutely overjoyed at all the Platinum Jubilee events, which also really brought our community together. I look back fondly on the Golden Jubilee too. I was 15 and I went to a street party where I used to live. It was the first time I ever drank alcohol (unbeknownst to my parents!). I definitely felt a sense of loss when The Queen died. She had always been there, serving selflessly and providing a sense of stability.'

74

Eileen
United Kingdom

'In 1953 I remember going to a party to celebrate The Queen's Coronation when I was about 8 years old. I was very emotional when I heard she had passed away, and it felt like a shock even though she was very old. I think Charles makes a wonderful King and is a worthy successor to the Crown.'

75

Rhys Addison
Student, United Kingdom

'We had the news on in the background but I wasn't paying full attention as I was scrolling through my phone, and then the television went quiet. I looked and the BBC logo had come up on a grey screen. I knew something was wrong and I called my mum in from the kitchen. It just didn't feel real to me. It felt like the world went silent after it was announced. We just watched in shock.'

76

Victoria Hedges
Plymouth, England

'I was at home, reading on the sofa. I instantly realised when they said she was unwell that this was history in the making. Not seen for decades and we were witness to it. The montages of her life made me feel very emotional. I felt very inspired after seeing her pure dedication. And very proud of her achievements.'

77

Craig Simpson
London, England

'Received a message on WhatsApp, then we put on the office television. The entire office stayed late to see what was going on. Then we went to the pub. It was very emotional after the announcement, and I remember seeing all the bus stops and advertisements showing The Queen's image in black and white.'

78

Anonymous
London, England

'I was out shopping when another customer said, "The children are going to Balmoral," and there was a muffled groan from the other shoppers. The assistant who served me was in tears and there was no conversation anywhere as people pulled out their phones to check the newsflash. Many simply left in silence.'

79

Joseph Robinson
Berkshire, England

'With her passing, the nation has lost an integral piece of itself; a burning light that has shone and warmed me all my life; a spark that had the capacity to ignite within me unbridled pride, joy and hope.'

80

Susan Black
Sheffield, England

'I was so sad when I heard The Queen had died. I thought she was a wonderful wise lady and held the country together. I still miss seeing her on the television with her lovely smile. My earliest memory of The Queen was her Coronation when I was only five years old.'

81

Dale Hurst
Bournemouth, England

'It was an ordinary day at work. One minute nothing, and the next, the entire university marketing team was gathering around a computer monitoring BBC News and Twitter, as the situation started to seem serious. Some of us were debating how our university should respond if The Queen died, others just talked about the idea of losing "The Nation's Nana".'

82

James Strickland
Carlisle, England

'I suppose we all knew it would come "one day" yet nonetheless the Queen's death came as a shock. Like so many, I felt that the one constant throughout my life was no longer. She wasn't just a leader, she really represented what was best about Britain. I will always be proud that most of my life was lived through her remarkable reign.'

83

Georgi Black
Fife, Scotland

'She had been around for so long, and as silly as it sounds, I assumed she would be around forever.'

84

Alex Shaw
Upminster, England

'I'm a train driver, and on the day of her death I got in from work and had turned the television on but had fallen asleep. My first realisation that she had died was opening my eyes and seeing 'EIIR' on the screen with the dates of her life underneath.

I admit I did shed a tear or two at the time. It felt like a great privilege to be living through history unfolding. The conclusion of a 70-year reign that I know I'll never see again.'

Follow-up question: Did you feel proud to be ferrying people to the Lying in State?

'That was very much how it felt; there were many people travelling and although sombre there was quite an atmosphere of joining-together amongst the travellers.

It was a pleasure to see people looking out for each other. I think it was an event that affected a lot of people more than they realised, expected, or were possibly prepared to admit. I've lived through the death of parents and other close relatives and family friends, each one signifying change and the ending of an era. The Queen was all of those, but also the end of national and to a great degree international continuity and stability as well.'

85

Carol Hawkins
Lincoln, England

'We were at home when we heard the news The Queen had passed away.

Hearing The Queen had passed away left an overwhelming deep sadness. Sharing the grief with my husband was a relief as we could hug and hold each other while we cried.

The Queen, our Monarch, the Mother of our Nation, had always been in my life. This may be a ridiculous thought, but I never imagined her not being there.

On the 9 September we, along with four of our friends, were due to attend a Battle of Britain Ball, which was of course cancelled.

So the six of us still met up for Chinese. It meant a lot to share our grief with dear friends. Then of course we had our traditional toast of port to Her Majesty Queen Elizabeth II.

We have a bar in our garden where we socialise with many of our friends who are veterans. We have always had a picture of The Queen as it is our tradition to raise a glass of port to her and to absent friends. We now have a picture of King Charles, although we do still have a picture up of The Late Queen where we shall carry on this tradition.'

86

Kyrenia Boucher
Horncastle, England

'The Queen sadly passed away on my wedding day, and once we all heard the news we held a two-minute silence in her honour.

What is also kind of nice was the song we played as we walked back down the aisle as husband and wife . . . 'You're My Best Friend' by QUEEN.'

87

Cindy Campbell McConnell
Carmel, Indiana
United States of America

'I was working on a quilt project in the sewing room with my machine facing the television. The programme was interrupted by a special report, and I knew it before it was announced. I bowed my head, said a prayer, and turned the machine off. No more quilt work for me that day.'

88

Jamie West
London, England

'"When it was all officially confirmed, we all ran to the front of the theatre to watch the lights being turned off at 7.00 p.m.'

89

Stanton Stephens
Ludlow, England

'I was in the middle of my aquafit class at Ludlow Leisure Centre, when the instructor stopped the music and told us right there in the pool that The Queen had died. It was a very solemn moment, until the music was put back on and we all carried on with the class! I felt very strange dancing around in the pool having just heard something so momentous and grave. When I changed afterwards and walked home, the news weighed heavily on my shoulders.

It was a dark evening and the mourning bell was already tolling from the tower of Ludlow Church. I remember the muffled tone of the bell because I had never heard that sound before.'

90

Michael Iddon
Taunton, England

'I found it very moving when The Queen passed. She will never be replaced.'

91

David Walker-Smith
Leicester, England

'I was brought up as a royalist. Every big total event we would sit as a family with neighbours, not moving for the whole day – sandwiches, flasks of tea and coffee – in silence, apart from the oohing and ahhing!

Spin forward 50 years The Queen was the constant in our lives, the reassurance, the magic, the unique, the majestic and the matriarch of our world.

That Thursday, my BBC News alert buzzed. I just knew. I didn't move from the television, watching the screen showing the closed gates of Balmoral with the very occasional comings and goings. As the news continued, my phone began to ring, friends, family, all saying the same thing. "She was invincible, we thought she would go on forever."

The death announcement came just as I was going out to a neighbour's wedding party. It took all my strength to go. Having lost my mum and dad during Covid, that feeling of loss, grief, came back in abundance. It's all we spoke about at the party.

Then I was rooted to the television and the sofa for a week. I hardly moved, watching it all unfold. The magic and the majestic continued – it was a whirlwind of British history in the making. But that sense of loss was huge. Her voice, her calmness, her words would be no more.'

92

Ken Man
London, England

'The day began as normal – just another busy day at the office in central London.

Around lunchtime my phone alerted me to the news that "The Queen was under medical supervision" at Balmoral. A colleague, also shocked at the breaking news alert, announced loudly "The Queen is unwell."

Everyone in the office began to wonder if it was finally time, speculating whether it had already happened and analysing The Queen's photograph from earlier in the week.

The day came towards a close, when the same colleague as earlier gasped and shouted out, "She has died!". The entire office began to whisper or gasp at the news. The atmosphere in the office became still, silent and shocked. We didn't know what to do. It was as if we had under-prepared and there was no manual to follow.

It sounds silly, but it had never crossed my mind that she would one day pass away.

As I left the office, the streets in central London felt eerily heavy and far too silent for rush hour. Everyone like me must have felt lost or unsure how to react to this shocking news. It hadn't sunk in yet.

A few days later I ended up joining 'the Queue'. Surprisingly it wasn't planned, and I joined on the Saturday afternoon with some friends.

My friends and I in a last-second decision decided to go and pay our respects to The Late Queen. We met in Southwark Park to join 'the Queue', with picnic items in hand, warm clothing and comfortable shoes. Once we received our special entry wristbands, we were ready for the long journey into the night.

'The Queue' became like a pilgrimage. I began to reminisce about The Queen during my own lifetime and my earliest memories seemed to be the funeral of Diana, Princess of Wales, and celebration of The Queen's Golden Jubilee. My own school had a Jubilee fete, and of course I recall seeing her every year on Christmas Day – no doubt it would be 3.00 p.m. sharp!

In a way it was a joyous queue, with lots of camaraderie and stories. I met people from all corners of the world while shuffling along the Thames.

There were my two friends from Surrey, a woman who stayed up for nearly 48 hours from Canada, another woman who decided to join by herself from Latvia, and a mother and daughter from Cambridge – each one joining for their own individual reason.

'The Queue' travelled through south London, along the River Thames and meandered past many famous monuments – many of which The Queen would have known well or opened.

The mood of 'the Queue' changed as we finally edged closer to Westminster Hall. There was a final zigzag round what seemed to be miles of airport-style barriers. Then there was security, only to finally enter Westminster Hall at 1.00 a.m. on Sunday morning.

It became silent apart from the shuffling feet and you could feel the presence of the guards.

The Late Queen was in front of me, Lying in State. It didn't feel real, and it was unimaginable to think The Queen could pass away.

I whispered a little message to her as I stared at the coffin, gave one last bow, and said, "Night night, rest well, Your Majesty."'

93

Leslie Jamieson
Toronto, Ontario, Canada

'Being a proud Canadian of a certain age, I remember well The Queen's portrait hanging in my schoolroom as a child. From a young age I always felt proud that The Queen belonged to and represented us as Canadians.

She was Queen of Canada and a human link between our past and future.

Much later in life I was very fortunate to be invited to an official event in Ottawa that she and Prince Philip were attending. Naturally it was one of the proudest moments of my life to be presented to her that day. The Queen's human warmth was something I will never forget.

I heard about her death when I was on a train in Toronto. A rather stunned silence went around after a fellow passenger blurted out, "Oh my, The Queen has died."

Then and there it felt like a chapter of history had closed, and a new one was beginning with the reign of a new King of Canada.'

94

Helen Simpson
Liverpool, England

'When I heard of The Queen's passing I was at work for the NHS. I really didn't take it in at the time, but when I got home I watched all the news from around the world documenting The Queen's passing. It didn't feel real that she was gone as she was a solid monarch for the country. The day of the State Funeral I watched the procession and service. There was a point when the coffin passed her horse, and the horse bowed – I cried.'

95

Kate Wilson
Buckinghamshire, England

'I was at my sister's home in Manchester on the day that The Queen died. We were glued to the news as the information about her being poorly and family arriving rapidly to Scotland – we knew it wasn't good. Then the news that she had passed. It was totally heartbreaking. We all shed a tear, my niece especially so, as it reminded her of her nana who was the same age as The Queen but had passed a few years earlier. So many memories of how it felt for them, and they were upset for The Queen's family.

I worked the day of the funeral [as a paramedic]. It was massive for me. I was on The Long Walk in Windsor, so it was quite busy and a very long day, but I felt privileged to be a part of it and to be able to say my own personal goodbye to The Queen.'

96

David Gillin
Liverpool, England

'At about lunchtime in the primary school where I teach, there were murmurs that The Queen was unwell. There had been rumours swirling around about her health for weeks, but the atmosphere in school felt different today. The staff in our school represent a diverse range of viewpoints about the monarchy, from staunch royalists (one of whom had even been invited to one of The Queen's garden parties), to republicans whose opinions of the Royal Family were . . . colourful, to say the least. Everyone, though, seemed just a little more sensitive to the news today.

When I led my Year 2 children in prayer at the end of the day (it's a Church of England school), praying for her health, her children, her grandchildren, her great-grandchildren, her horses and her corgis, I had never seen their little eyes scrunched so tight and their hands clenched so earnestly.

Early that evening, it was confirmed. As one of the few teachers trusted with the password to our school Twitter account, it was down to me to post something suitably poignant, meaningful and sensitive to mark the occasion. Fifteen redrafts later, and after frantic messages to and from our senior leadership team, a simple portrait with a black background was decided upon. It was all that was needed. No words. A little later, the Church of England posted a prayer that we shared and then finally, a retweet from Paddington Bear, simply saying, "Thank you, Ma'am, for everything."

Later when her time of death – 3.10 p.m. – was announced, it struck me that it would have been the exact time that my little six- and seven-year-olds were sitting, five minutes before home time, coats on and rucksacks at the ready, in prayer. All 30 of them. Hands together, eyes closed, heads bowed, hearts open. Maybe one day, they'll remember that. They won't remember me, though, smiling softly and sadly, watching their little innocent faces as – unbeknownst to us – our country changed forever.'

96
Quotes by Queen Elizabeth II

1

'I declare before you all that my whole life, whether it be long or short, shall be devoted to your service and the service of our great imperial family to which we all belong.'

The Queen's 21st birthday speech – 21 April 1947

2

'Forgiveness lies at the heart of the Christian faith. It can heal broken families, it can restore friendships and it can reconcile divided communities. It is in forgiveness that we feel the power of God's love.'

From The Queen's 2011 Christmas Broadcast

3

'I cannot lead you into battle. I do not give you laws or administer justice. But I can do something else. I can give you my heart and my devotion to these old islands and to all the peoples of our brotherhood of nations.'

First televised Christmas Broadcast in 1957

4

'What I say to you now, as your Queen and as a grandmother, I say from my heart . . . No one who knew Diana will ever forget her. Millions of others who never met her, but felt they knew her, will remember her.

'I for one believe there are lessons to be drawn from her life and from the extraordinary and moving reaction to her death. I share in your determination to cherish her memory.'

The Queen's televised address following the death of Diana, Princess of Wales in September 1997

5

'It does matter therefore what each individual does each day. Kindness, sympathy, resolution and courteous behaviour are infectious.'

From The Queen's Christmas Broadcast in 1975

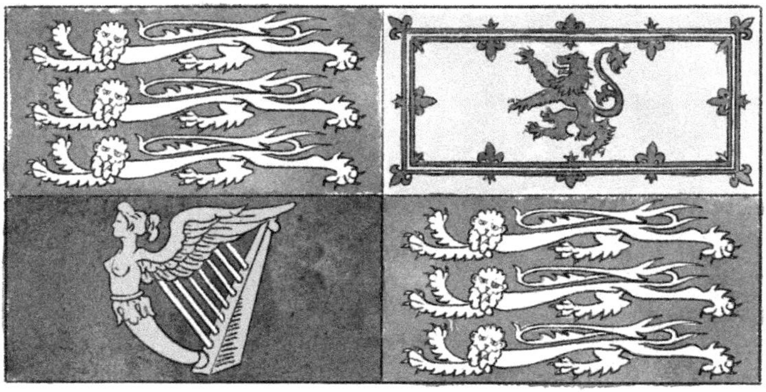

6

'Like all the best families, we have our share of eccentricities, of impetuous and wayward youngsters and of disagreements.'

The Queen in October 1989

7

'The struggles against inhuman prejudice, against squalor, ignorance and disease, have always owed a great deal to the determination and tenacity of women.'

From The Queen's 1966 Christmas Broadcast

8

'If I am asked today what I think about family life after 25 years of marriage, I can answer with equal simplicity and conviction. I am for it.'

The Queen's Silver Wedding anniversary speech, 1972

9

'We cheered The King and The Queen on the balcony and then walked miles through the streets. I remember lines of unknown people linking arms and walking down Whitehall, all of us just swept along on a tide of happiness and relief.

'I think it was one of the most memorable nights of my life.'

Speaking to the BBC in 1985 about joining street celebrations for VE Day in London, May 1945

10

'You children have something to give us which is priceless. You can still look at the world with a sense of wonder and remind us grown-ups that life is wonderful and precious.'

From The Queen's 1989 Christmas Broadcast

11

'He is someone who doesn't take easily to compliments but he has, quite simply, been my strength and stay all these years, and I, and his whole family, and this and many other countries, owe him a debt greater than he would ever claim, or we shall ever know.'

The Queen's tribute to The Duke of Edinburgh during their Golden Wedding anniversary celebrations in 1997

12

'I often draw strength from meeting ordinary people doing extraordinary things: volunteers, carers, community organisers and good neighbours; unsung heroes whose quiet dedication makes them special.

'They are an inspiration to those who know them.'

From The Queen's 2016 Christmas Broadcast

13

'While we have faced challenges before, this one is different. This time we join with all nations across the globe in a common endeavour, using the great advances of science and our instinctive compassion to heal. We will succeed – and that success will belong to every one of us.

'We should take comfort that while we may have more still to endure, better days will return: we will be with our friends again; we will be with our families again; we will meet again.'

The Queen's televised address on 5 April 2020
during the COVID-19 pandemic

14

'Although the contribution which any one person can make is small, it is real and important.'

From The Queen's Christmas Broadcast in 1960

15

'Grief is the price we pay for love.'

Following the 11 September 2001 terrorist attacks in the United States

16

'To all those who have suffered as a consequence of our troubled past I extend my sincere thoughts and deep sympathy. With the benefit of historical hindsight we can all see things which we would wish had been done differently or not at all.'

A speech during The Queen's State Visit to Ireland in 2011

17

'Once again we are reminded that people matter, and it is our relationship with one another that is most important.'

From The Queen's 1975 Christmas Broadcast

18

'It's all to do with the training: you can do a lot if you're properly trained.'

The Queen speaking during the BBC documentary Elizabeth R *in 1992*

19

'I have in sincerity pledged myself to your service, as so many of you are pledged to mine. Throughout all my life and with all my heart I shall strive to be worthy of your trust.'

The Queen on her Coronation Day, 2 June 1953

20

'But let us not take ourselves too seriously. None of us has a monopoly of wisdom and we must always be ready to listen and respect other points of view.'

From The Queen's Christmas Broadcast in 1991

21

'Each day is a new beginning, I know that the only way to live my life is to try to do what is right, to take the long view, to give of my best in all that the day brings, and to put my trust in God.'

From The Queen's 2002 Christmas Broadcast

22

'The Duke of Edinburgh has made an invaluable contribution to my life over these past fifty years, as he has to so many charities and organisations with which he has been involved.'

The Queen speaking at Guildhall, London, 4 June 2002

23

'Over the years, those who have seemed to me to be the most happy, contented and fulfilled have always been the people who have lived the most outgoing and unselfish lives.'

From The Queen's 2008 Christmas Broadcast

24

'Inevitably, a long life can pass by many milestones; my own is no exception.'

The Queen in 2015

25

'It has always been easy to hate and destroy. To build and to cherish is much more difficult.'

From The Queen's 1957 Christmas Broadcast

26

'Such a blend of traditions serves to make us stronger, individually and collectively, by providing the ingredients needed for social, political and economic resilience.'

The Queen speaking about The Commonwealth in 2013

27

'I know of no single formula for success, but over the years I have observed that some attributes of leadership are universal, and are often about finding ways of encouraging people to combine their efforts, their talents, their insights, their enthusiasm and their inspiration, to work together.'

The Queen addressing The United Nations General Assembly in 2010

28

'We all need to get the balance right between action and reflection. With so many distractions, it is easy to forget to pause and take stock.'

From The Queen's 2013 Christmas Broadcast

29

'We know the reward is peace on earth, goodwill toward men, but we cannot win it without determination and concerted effort.'

From The Queen's 1963 Christmas Broadcast

30

'It is through this lens of history that we should view the conflicts of today, and so give us hope for tomorrow.'

From The Queen's 2011 Christmas Broadcast

31

'Our modern world places such heavy demands on our time and attention that the need to remember our responsibilities to others is greater than ever.'

From The Queen's 2002 Christmas Broadcast

32

'This Assembly was born of the endeavours of countless men and women from different nations who, over the centuries, have pursued the aims of the preservation of peace between nations, equality of justice for all before the law and the right of the peoples of the world to live their lives in freedom and security.

The Queen addressing the United Nations General Assembly in 1957

33

'His [Prince Philip's] sense of service, intellectual curiosity and capacity to squeeze fun out of any situation were all irrepressible . . . That mischievous, enquiring twinkle was as bright at the end as when I first set eyes on him.'

From The Queen's final Christmas Broadcast in 2021

34

'Reflecting on these events makes me grateful for the blessings of home and family and, in particular, for 70 years of marriage. I don't know that anyone had invented the term "platinum" for a 70th wedding anniversary. When I was born, you weren't expected to be around that long.'

From The Queen's 2017 Christmas Broadcast

35

'Therefore I am sure that this, my Coronation, is not the symbol of a power and a splendour that are gone but a declaration of our hopes for the future, and for the years I may, by God's Grace and Mercy, be given to reign and serve you as your Queen.'

The Queen on her Coronation Day, 2 June 1953

36

'We are reminded here [in Parliament] of our past, of the continuity of our national story and the virtues of resilience, ingenuity and tolerance which created it.'

The Queen's Diamond Jubilee address to Parliament, 2012

37

'There is no point in regretting the passage of time. Growing older is one of the facts of life, and it has its own compensations. Experience should help us to take a more balanced view of events and to be more understanding about the foibles of human nature.'

From The Queen's 1987 Christmas Broadcast

38

'When I was 21, I pledged my life to the service of our people and I asked for God's help to make good that vow. Although that vow was made in my salad days, when I was green in judgement, I do not regret nor retract one word of it.'

The Queen's speech at Guildhall, London,
to celebrate her Silver Jubilee in 1977

39

'Gratitude, respect and pride, these words sum up how I feel about the people of this country and the Commonwealth – and what this Golden Jubilee means to me.'

The Queen in 2002 during her Golden Jubilee celebrations

40

'Everyone is our neighbour, no matter what race, creed or colour.'

From The Queen's 2004 Christmas Broadcast

41

'Even when your life seems most monotonous, what you do is always of real value and importance to your fellow men.'

From The Queen's 1954 Christmas Broadcast

42

'It's inevitable that I should seem a rather remote figure to many of you, a successor to the kings and queens of history.'

From The Queen's 1954 Christmas Broadcast

43

'The path, of course, is not always smooth, and may at times this year have felt quite bumpy, but small steps can make a world of difference.'

From The Queen's 2019 Christmas Broadcast

44

'Whatever life throws at us, our individual responses will be all the stronger for working together and sharing the load.'

The Queen speaking in Ireland in 2011

45

'She had an infectious zest for living, and this remained with her until the very end. I know too that her faith was always a great strength to her.'

The Queen reflecting on the death of Queen Elizabeth The Queen Mother in 2002

46

'In remembering the appalling suffering of war on both sides, we recognise how precious is the peace we have built in Europe since 1945.'

The Queen in 2004

47

'Though we each lead different lives, the experience of growing older, and the joys and emotions which it brings, are familiar to us all.'

From The Queen's Christmas Broadcast in 1998

48

'Being united – that is, feeling a unity of purpose – is the glue that bonds together the members of a family, a country, a Commonwealth. Without it, the parts are only fragments of a whole; with it, we can be much more than the sum of those fragments.'

From The Queen's 1997 Christmas Broadcast

49

'We may hold different points of view but it is in times of stress and difficulty that we most need to remember that we have much more in common than there is dividing us.'

From The Queen's 1974 Christmas Broadcast

50

'I speak of a tolerance that is not indifference, but is rather a willingness to recognise the possibility of right in others; of a comradeship that is not just a sentimental memory of good days past, but the certainty that the tried and staunch friends of yesterday are still in truth the same people today; of a love that can rise above anger and is ready to forgive.'

From The Queen's 1956 Christmas Broadcast

51

'We are a moderate, pragmatic people, more comfortable with practice than theory.'

Reflecting on Britain in 2002

52

'The true measure of all our actions is how long the good in them lasts . . . everything we do, we do for the young.'

The Queen in 2014

53

'The ceremonies you have seen today are ancient, and some of their origins are veiled in the mists of the past. But their spirit and their meaning shine through the ages never, perhaps, more brightly than now.'

The Queen on her Coronation Day, 2 June 1953

54

'I have lived long enough to know that things never remain quite the same for very long.'

From The Queen's 2006 Christmas Broadcast

55

'In an era when the regular worthy rhythm of life is less eye-catching than doing something extraordinary, I am reassured that I am merely the second sovereign to celebrate a Diamond Jubilee.'

The Queen in 2012

56

'No matter what scientific progress we make, the message will count for nothing unless we can achieve real peace and encourage genuine goodwill between people and the nations of the world.'

From The Queen's Christmas Broadcast in 1967

57

'I believe in our qualities and in our strength, I believe that together we can set an example to the world which will encourage upright people everywhere.'

From The Queen's Christmas Broadcast in 1957

58

'At Christmas I am always struck by how the spirit of togetherness lies also at the heart of the Christmas story.'

From The Queen's Christmas Broadcast in 1958

59

'None of us underestimates the challenges ahead: but history has shown that when nations come together in common cause, there is always room for hope. Working side by side, we have the ability to solve the most insurmountable problems and to triumph over the greatest of adversities.'

The Queen's address to the COP26
United Nations Climate Change Summit, 2021

60

'As we all look forward to the start of a new decade, it's worth remembering that it is often the small steps, not the giant leaps, that bring about the most lasting change.'

From The Queen's 2019 Christmas Broadcast

61

'Goodwill towards men" is not a hollow phrase. Goodwill exists, and when there is an opportunity to show it in practical form we know what wonderful things it can achieve.'

From The Queen's 1965 Christmas Broadcast

62

'I hope that Christmas will remind us all that it is not how we communicate but what we communicate with each other that really matters.'

From The Queen's 1983 Christmas Broadcast

63

'I am constantly amazed by the generosity of donors and subscribers, great and small, who give so willingly and often towards the enjoyment of others.'

From The Queen's 1991 Christmas Broadcast

64

'This message – love thy neighbour as thyself – may be for Christians 2,000 years old. But it is as relevant today as it ever was. I believe it gives us the guidance and the reassurance we need as we step over the threshold into the twenty-first century.

'And I for one am looking forward to this new millennium.'

From The Queen's 1999 Christmas Broadcast

65

'When peace comes, remember it will be for us, the children of today, to make the world of tomorrow a better and happier place.'

Wartime radio broadcast alongside Princess Margaret, 13 October 1940

66

'There may be an instinct in all of us to help those in distress, but in many cases I believe this has been inspired by religious faith. Christianity is not the only religion to teach its followers to help others and to treat your neighbour as you would want to be treated yourself.'

From The Queen's 2005 Christmas Broadcast

67

'The Christmas message to each of us is indivisible; there can be no "Peace on earth" without "Goodwill toward men".'

From The Queen's 1955 Christmas Broadcast

68

'How is it, people wonder, that we are forever seeking new worlds to conquer before we have properly put our own house in order.'

From The Queen's 1962 Christmas Broadcast

69

'A lack of humanity and compassion can be very destructive – how easily this causes diversions within nations and between nations. We should remember instead how much we have in common and resolve to give expression to the best of our human qualities, not only at Christmas, but right through the year.'

From The Queen's 1973 Christmas Broadcast

70

'Sooner or later we all become aware of the passing of the years, but every now and then we get a sharp reminder that time is moving on rather quicker than we expected.'

From The Queen's 1987 Christmas Broadcast

71

'Learning to be concerned about one another; to treat your neighbour as you would like him to treat you; and to care about the future of all life on earth.

'These matters of the spirit are more important and more lasting than simple material development.'

From The Queen's 1970 Christmas Broadcast

72

'There is a lesson in this for us all and we should never forget our obligation to make our own individual contributions, however small, towards the sum of human goodness.'

From The Queen's 1985 Christmas Broadcast

73

'As human beings we generally know what is right and how we should act and speak. But we are also very aware of how difficult it is to have the courage of our convictions.'

From The Queen's 1981 Christmas Broadcast

74

'My father, and my grandfather before him, worked all their lives to unite our peoples ever more closely, and to maintain its ideals which were so near to their hearts. I shall strive to carry on their work.'

From The Queen's first Christmas Broadcast, 1952

75

'You will be keeping it [Coronation Day] as a holiday; but I want to ask you all, whatever your religion may be, to pray for me on that day – to pray that God may give me wisdom and strength to carry out the solemn promises I shall be making, and that I may faithfully serve Him and you, all the days of my life.'

From The Queen's first Christmas Broadcast, 1952

76

'At the start, the outlook seemed bleak, the end distant and the outcome uncertain, but we kept faith that the cause was right. And this belief as my father noted in his broadcast carried us through. "Never give up. Never despair." That was the message of VE Day.'

From The Queen's broadcast on 8 May 2020, the 75th anniversary of VE Day

77

'There is a motto which has been borne by many of my ancestors – a noble motto, "I serve".'

Princess Elizabeth in 1947

78

'If you throw a stone into a pool, the ripples go on spreading outwards. A big stone can cause waves, but even the smallest pebble changes the whole pattern of the water. Our daily actions are like those ripples, each one makes a difference, even the smallest.'

From The Queen's 1975 Christmas Broadcast

79

'All life on earth depends on how we behave towards one another, and how we treat the plants and the animals that share our world with us.'

From The Queen's 1989 Christmas Broadcast

80

'Prince Philip is, I believe, well-known for declining compliments of any kind. But throughout he has been a constant strength and guide.'

The Queen's address to Parliament on her Diamond Jubilee, 2012

81

'It is particularly at Christmas, which marks the birth of the Prince of Peace, that we should work to heal old wounds and to abandon prejudice and suspicion.'

From The Queen's 1984 Christmas Broadcast

82

'Even if the problems seem overwhelming, there is always room for optimism. Every problem presents us with the opportunity both to find an answer for ourselves and to help others.'

From The Queen's 1978 Christmas Broadcast

83

'Perhaps this Jubilee is a time to remind ourselves of the benefits which union has conferred, at home and in our international dealings, on the inhabitants of all parts of this United Kingdom.'

The Queen's address to Parliament on her Silver Jubilee, 1977

84

'The most gracious of men [South African President Nelson Mandela] has shown us all how to accept the facts of the past without bitterness, how to see new opportunities as more important than old disputes and how to look forward with courage and optimism.

'His example is a continuing inspiration to the whole Commonwealth and to all those everywhere who work for peace and reconciliation.'

From The Queen's 1996 Christmas Broadcast

85

'I believe that strong and open communities matter both in good times as well as bad.'

From The Queen's 2001 Christmas Broadcast

86

'Christ's example has taught me to seek to respect and value all people of whatever faith or none.'

From The Queen's 2014 Christmas Broadcast

87

'At the same time I want to show that the Crown is not merely an abstract symbol of our unity but a personal and living bond between you and me.'

From The Queen's 1953 Christmas Broadcast

88

'I come across examples of unselfish service in all walks of life and in many unexpected places.'

From The Queen's 1980 Christmas Broadcast

89

'The gift I would most value next year [1977] is that reconciliation should be found wherever it is needed. A reconciliation which would bring peace and security to families and neighbours at present suffering and torn apart.'

From The Queen's 1976 Christmas Broadcast

90

'On our own, we cannot end wars or wipe out injustice, but the cumulative impact of thousands of small acts of goodness can be bigger than we imagine.'

From The Queen's 2016 Christmas Broadcast

91

'Some cultures believe a long life brings wisdom. I'd like to think so.'

From The Queen's 2018 Christmas Broadcast

92

'For more than seventy years, I have been lucky to meet and to know many of the world's great leaders. And I have perhaps come to understand a little about what made them special.

'It has sometimes been observed that what leaders do for their people today is government and politics. But what they do for the people of tomorrow – that is statesmanship.'

The Queen's address to the COP26 United Nations Climate Change Summit, 2021

93

'We think of our homes as places of warmth, familiarity and love; of shared stories and memories, which is perhaps why at this time of year so many return to where they grew up. There is a timeless simplicity to the pull of home.'

From The Queen's 2017 Christmas Broadcast

94

'I have to be seen to be believed.'

Attributed to The Queen when commenting on her bright-coloured wardrobe

95

'But through the many changes I have seen over the years, faith, family and friendship have been not only a constant for me but a source of personal comfort and reassurance.'

From The Queen's 2018 Christmas Broadcast

96

'But life, of course, consists of final partings as well as first meetings …'

From The Queen's final Christmas Broadcast in 2021

Acknowledgements

Compiling *96: What The Queen Meant To Us* was a project involving many people – more than 96 in fact. From the beginning my desire was to capture personal stories from as diverse a background as possible, from different parts of the United Kingdom, the Commonwealth and the wider world.

The project began with a concept and was taken to the next level by the investigative and journalistic research of AJ West, who sought out some of the participants and their stories.

Of course 96 could never have taken shape without the individual contributors to which we extend our gratitude for sharing their personal stories:

Adam Boulton
Adam Kerr
Al Jones
Alex Clement
Alyssa Lamanna
Amy B. Garratt
Andrew David Stedman
Andrew McDaniel
Barry Stewart
Charlie Morris
Charlotte from Berkshire
Dale Hurst
Danielle Amor
David Gillin
David Walker-Smith
Davina Woodhouse
Debbie Ioanna
Diane Clarke
Elle Padilla
Eloise from Nottingham

Emma Mortimer
Emma Nankervis
Emma Sommerville
Emma, student from Scotland
Fiona from Downpatrick
Fiona Brown
Frederick Hervey-Bathurst
Gail Popock
James from Dorset
Gareth Hamilton
Gary Simmons
Georgi Black
Alex Shaw
Carol Hawkins
Kyrenia Boucher
Cindy Campbell McConnell
Georgina Wright
Graham McAulay-Sillars
Hannah Catherall
Hassan Jain
Hayley Anderton
Heather Crosskey
Helen Simpson
Jacqueline Surridge
Jade Nomura
James Strickland
Jamie West
Jane Armstrong
Jemima Agyemang
Jemma Hole
Joanne Greensill
Joseph Robinson
Kate Barton
Kate Wilson
Kelly Holland
Lauren from Ayrshire
Lauren Sparks
Marie Hastings
Martyn Dore
Matthew from the West Midlands
Michael Iddon
Leslie Jamieson
Michelle Nicholls
Natasha Bayliss
Nicola Doyle
Nicola Smith
Nina Smith
Paul and Emily from North Norfolk
Peter Steer
Rhys Addison
Robert Bailey
Robynne Williams
Russell Deathridge
Saif Mokeef
Sally from the East Midlands
Samantha Thomas
Sarah Evans
Sean Weston
Stanton Stephens
Steph Barker
Stephanie Wadsworth
Stewart Davies
Stuart Cummins

Susan Black
Tim Jukes
Toni Broom
Victoria Hedges
Craig Simpson
Wendy Robey

Additionally I cannot thank Cyril Worthington-Feckenham enough for his in-depth research into the period between 8–19 September 2022. This involved countless enquiries with official government departments, news agencies and experts on Operation Unicorn and Operation London Bridge.

Special thanks as well to The London Library for providing me with a constant refuge for research and writing.

A personal thank you to the following people who were helpful with the editing, direction, and look and feel of the project: Ken Man, Jeffrey Hyland, Douglas Walker, Frances Prior-Reeves, Andrew Prescott, Karen Mangia, and Ken Stone.

Andrew Lannerd
London, Spring 2025